LAMBERTVILLE FREE PUBLIC LIBRARY
D0114511

How Old is Your House?

a guide to research

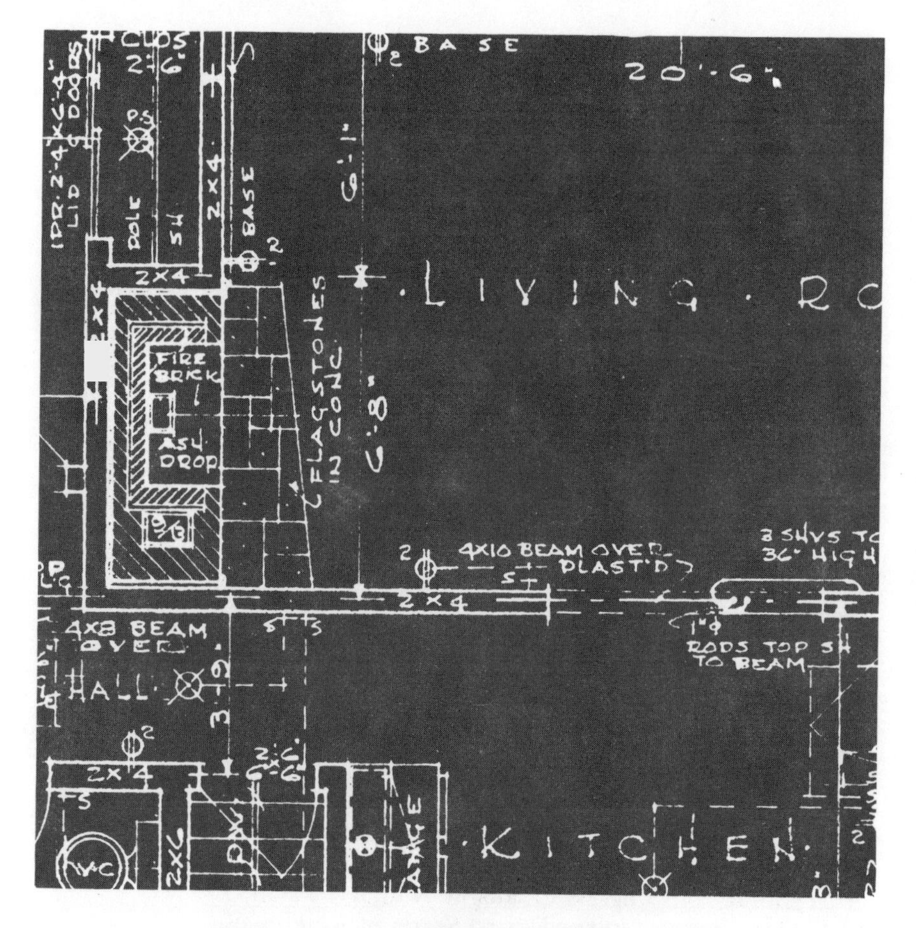

Joan Webber

Chester, Connecticut

ISBN: 0-87106-092-2
Library of Congress Catalogue Number: 77-88359
Manufactured in the United States of America

Third Printing

Table of Contents

I. INTRODUCTION 7

II. HOW DO YOU START? 9
Look at your house — Examine it carefully.

III. WHAT STYLE IS YOUR HOUSE? 19
Center Chimney Colonial — Salt Box — Cape Cod — Early Georgian — High Georgian — Federal — Greek Revival — Gothic Revival — Italianate — Mansard — Queen Anne.

IV. WHERE TO DIG FOR FACTS 65
Deeds and plans — Wills — Town History and Genealogy — Published and unpublished books, articles, memoires — Maps — Town Register — Historical Society — Assessors and Tax Collectors Offices — Books and articles on old houses.

V. WHAT DOES IT ALL MEAN? 83

VI. A CASE STUDY 84
The John Mason House.

VII. POST SCRIPT 93
Write your own report.

VIII. GLOSSARY 94
Deed terms.

IX. BIBLIOGRAPHY 100
List of old house books.

Acknowledgements

Margery Howard, Reference Supervisor at the Lexington Library, gave me the idea for this book, when I unwittingly joined a long list of others, to ask a nuisance question about researching my house. She cheerfully helped me then and has been delightful support since.

Anne Grady and Betsy Turvene took precious time from their own pursuits to read my initial manuscript and add helpful suggestions.

Elizabeth Reinhardt not only took an active interest and read the entire work, but made me toe the line and be more specific with many details. I am grateful for her aid and patience.

And, finally, Martha Mabee is my moral support and famous ally.

To Jim
For His Encouragement
and Understanding

Celebrations of the Bicentennial year have made us all more aware of our heritage and awakened desires to learn about our past and surroundings. What better place is there to understand than the house in which we live? For those who suffer the vicissitudes of an old house, a knowledge of its history and insight into the lives of its past inhabitants, can make the whole experience worth while.

John Mason House

How Old is Your House?
A Guide to Research

I INTRODUCTION

The age of old houses has long piqued the curiosity of many people. With a little work and perserverence, most of these houses can be dated fairly accurately and the step by step progression back in time is rewarding. As you uncover the saga of your home, you can not fail to become intrigued by tidbits of information on the lives of its former residents.

How many rooms did your house have originally? And how did it grow? The location and number of acres that surround it are important considerations. It also makes a difference if the early owners were farmers or tradesmen and whether they were active in their local church and government.

Although this guide is for houses of all ages, researching a house fifty to seventy five years old is quite a straight-forward task. More recent records are typed so that they are easy to read and they are recorded systematically, avoiding the loss of important documents. However, the search technique is the same as that of older houses, even if results are more quickly and easily obtained.

The pursuit of the origins of any house may be simple or complex, but either way the process can be delightful, with satisfying results. This guide will help save time by starting you in the right direction and it will be an aid past snags which are often encountered.

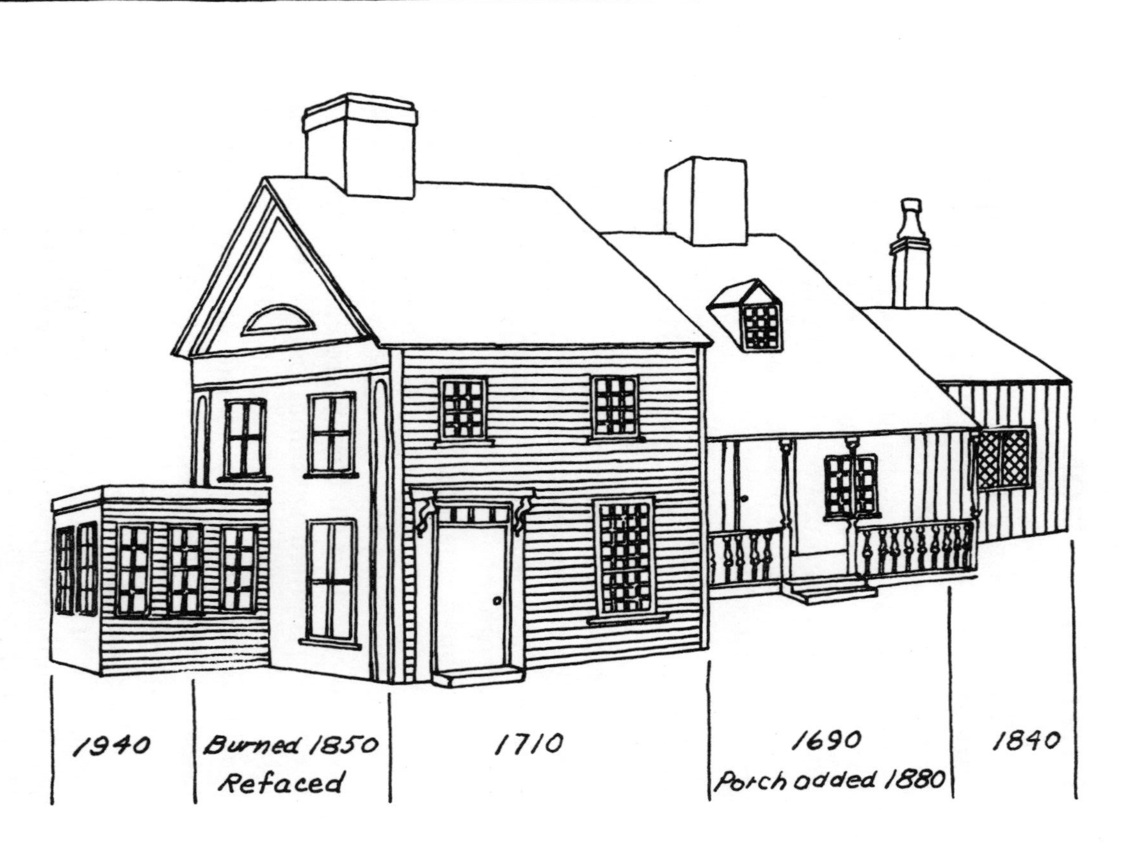

Problem House

II HOW DO YOU START?

Look at Your House

Nearly every house has a descernable style. The "vernacular" farm house is the most elusive, and if yours is of that category, it will be a watered down version of a specific architectural style or a practical farm dwelling with token stylistic touches of the era in which it was built. It may also reflect a confusing jumble of many periods that can result from cumulative changes of past owners. Otherwise, the style will range from early Colonial, Salt Box or Cape Cod through elegant Georgians and Federal to the more diverse later Gothic and Queen Anne periods.

What is the outside like? Overlapping clapboard ends and wavy window glass are an indication of age. Roof angles and raised roofs may tell you something. Do not be fooled by large window panes or ornamental brackets that hold up a porch roof. They could be authentic and in keeping with your period or add-ons. The many changes that may have been made to your house can disguise it, but they are only superficial additions of a more recent time. Strip them off mentally to reduce your home to its former self.

A clear-cut style is datable within a general time period, but you will have to narrow that time down considerably, because the span could be as much as fifty years or more. A "vernacular" farm house, of any design, also may have had a nebulous vogue in need of clarification.

"The added-on look"

Examine Your House Carefully

Study your home in detail. Become an old house detective and get acquainted with every inch, inside and out. Any clue may give a hint about its age.

Does your house face south? Early settlers built their houses facing the south for maximum benefit from the sun's warmth and light. With the phasing out of old roads and the building of new, these places may now front on the street, sit at an angle or even show the gable end.

Does it have the added-on look? In determining age you will want to decide which section is the oldest. Do not assume that the kitchen was added on and then the shed and barn, if you still have one. The kitchen could be an original one room house and when an early owner got enough money together, or his family expanded, he may have built on the larger two story part. A house can also be a composite of two small buildings, each of a different age.

Study the construction. Examine beams in your cellar and attic. They may show the marks of a broadax or of a circular saw. And they will be held in place by wooden pegs, hand made or machine made nails. Never miss an opportunity to look inside the walls if plaster should crumble and fall off or you have alterations and repairs made. You have to crawl on your hands and knees or slip on your back to see through crevices and crawl spaces. Use the old mirror trick. Shine a flashlight in a place too small to reach and hold a mirror behind the light where you can see the reflection. This is an excellent method for looking up chimneys and around corners.

"... slip on your back ..."

Dovetailed

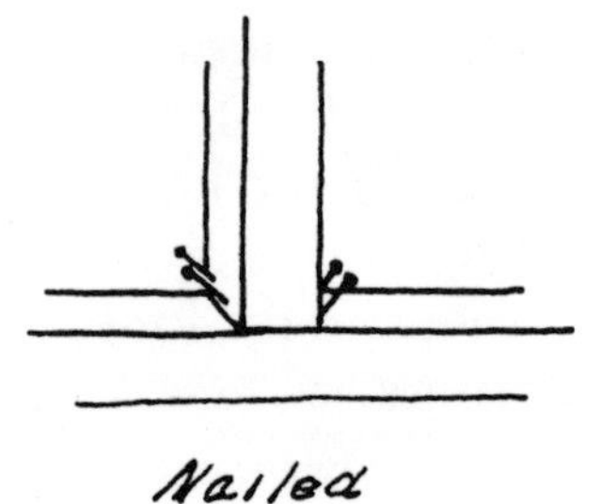

Nailed

Up and down

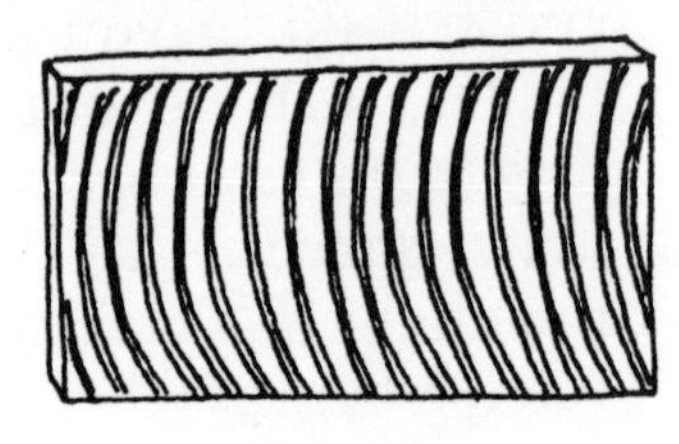

Circular

Saw marks

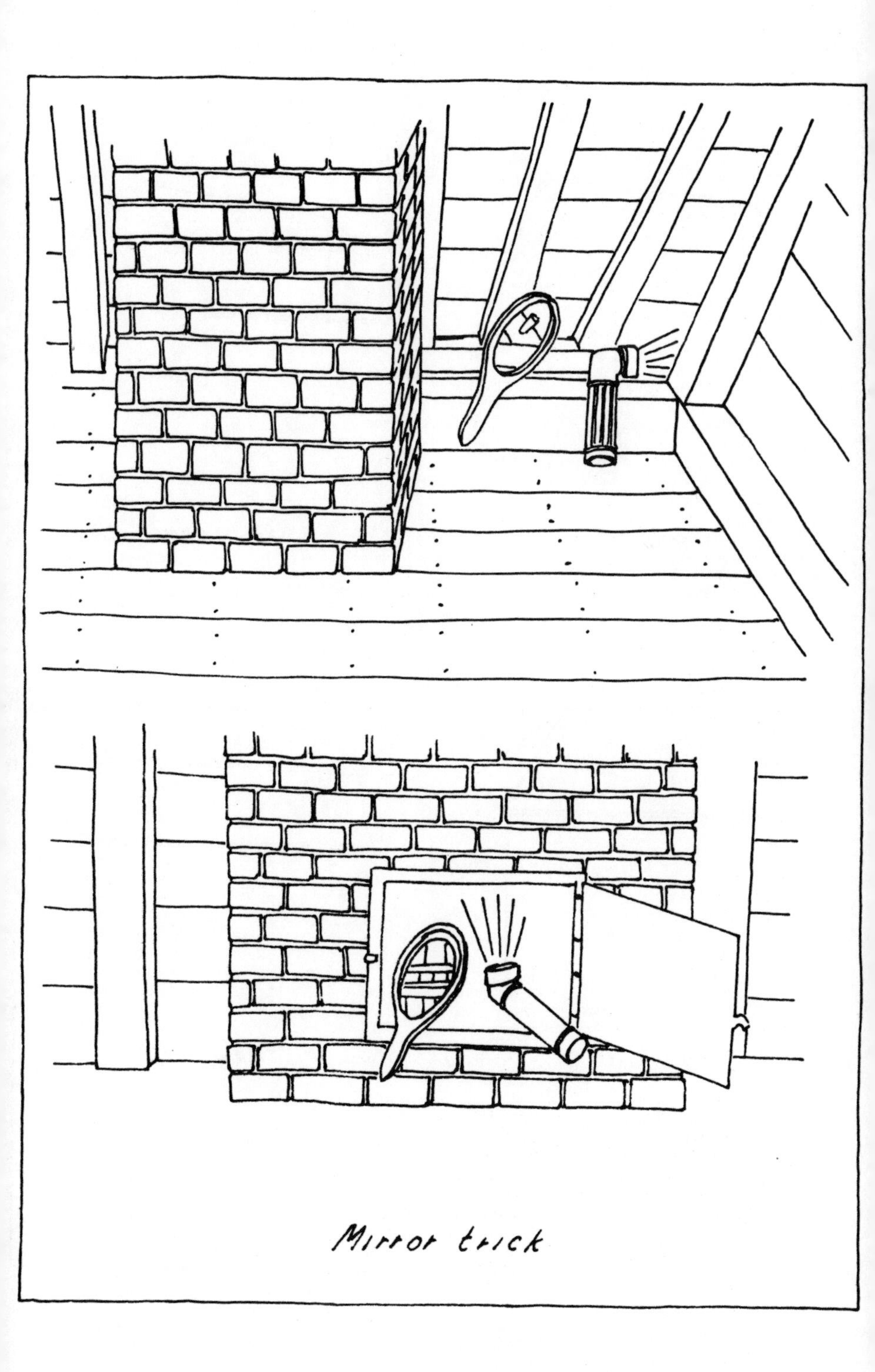

Mirror trick

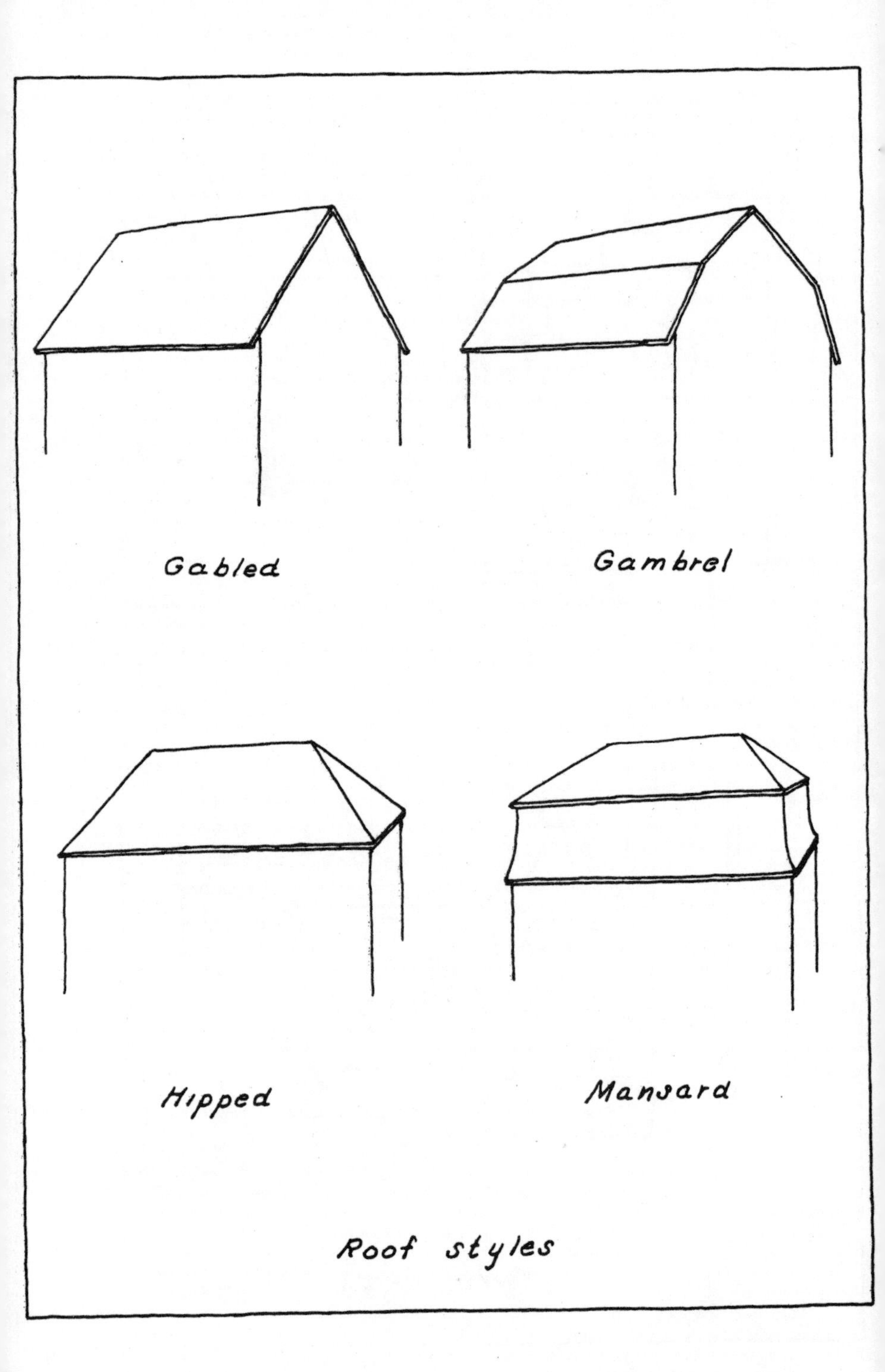

Roof styles

Some of your attic and roof construction might be original. Additions and roof changes are often visible, with more recent activity dramatically different from that of initial work. Be cautious, however, about assuming that rough rafters or boards are older than smoothed, neatly squared-off ones. Many roofs have been repaired or altered with logs still retaining their bark in sharp contrast to earlier more carefully dressed rafters.

A cellar can tell many stories. Check old brick work or sections of brick foundation and try to figure out whether they represent an original chimney. During the seventeen and eighteen hundreds, it was fashionable to take out a center chimney and replace it with two chimneys. Or perhaps a center chimney was converted and made smaller to balance an addition and second chimney.

Another word of caution. An ancient summer beam in your cellar that seems different from the rest of the construction, may have been recycled. It was not unusual for beams to be taken from a burned or abandoned place and used in the framework of a new building. Consider, also, that the farther you live from a metropolitan area, the more likely builders were to use the old methods of construction for a longer time. Brick and fieldstone foundations should be scrutinized. The original masonry was often replaced later by slabs of granite at ground level.

Can you determine where the old cellar stairs were or which area was used for the coal bin? Sometimes you can look up and see the number of layers of floor boards, although many floors have been replaced or the boards turned

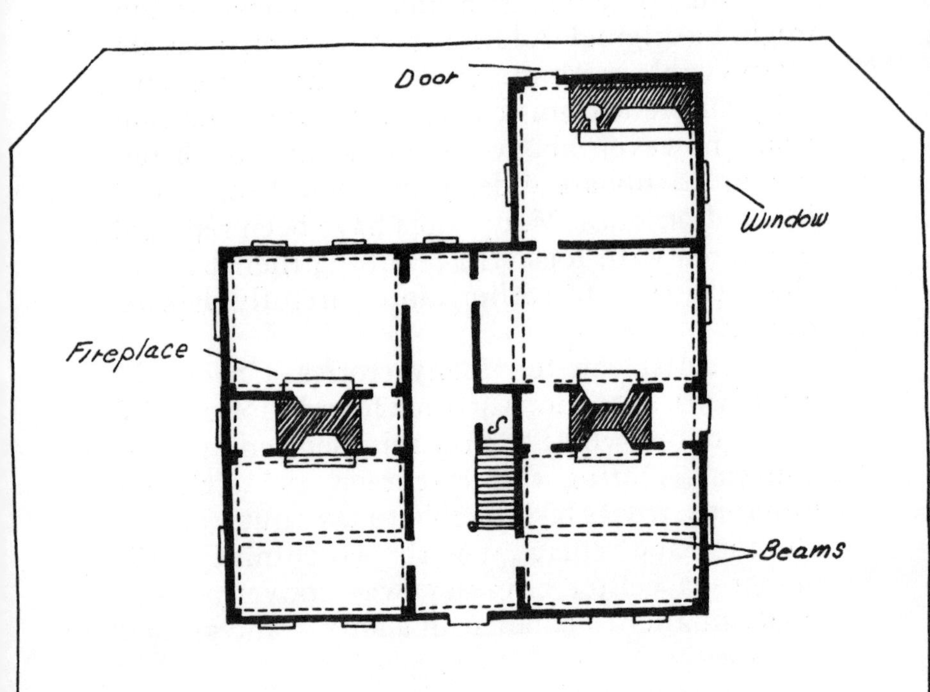

Ground floor

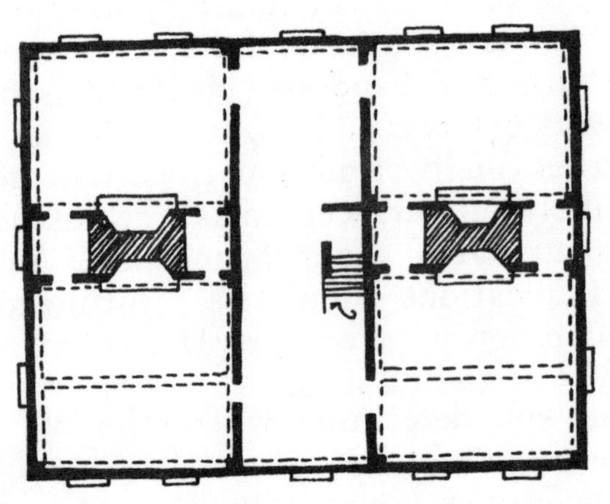

Second floor

Examples of floor plans

over and sanded. Smaller supporting beams will be mortised into larger ones with wooden pegs in earlier construction, butted and nailed later.

Study the placement of your rooms. You may have a bathroom where there was once a stairway or center chimney. Maybe a stairway replacing a pantry or storeroom. Earlier houses had the kitchen in what is perhaps your living room or dining room, while in later houses the kitchen may still be the kitchen. Some places retain a few pieces of original hardware, door latches and hinges or an isolated hook. These can help you identify the first room or rooms and their purpose.

Finally, try to draw a floor plan for each story and the basement. It will give you a clear picture of the exact location and relationship of each room to the whole of your house and will put chimneys and fireplaces in perspective. A clever method of doing this is to draw on heavy tracing paper and then superimpose the floors on each other. Any odd construction will become immediately apparent. Perhaps you will discover a hidden room or concealed storage space, crammed full of treasures!

The importance of a thorough knowledge of your house construction should not be underestimated, but it is equally essential to be wary of hasty judgements regarding the age of that construction.

"One room and loft"

III WHAT STYLE IS YOUR HOUSE?

Each of the twelve styles, to be more clearly delineated, has salient characteristics indicative of its period. Even in the most nondescript "vernacular", there is usually a proud detail proclaiming its right to be categorized. The following are typical descriptions, not all necessarily embodied in any given house, but some of which would have to appear in that house in order to justify its architectural placement.

Center Chimney Colonial

The enormously massive braced-frame of the Center Chimney Colonial distinguishes it as the "Grandaddy" of all building in this country. Its durability, a result of stalwart workmanship with crude tools, was not only expedient, but absolutely trustworthy.

This house replaced the temporary dwelling put up by settlers as they moved inland to farm the land. The earliest homes, having one or two rooms and loft, with a shed often added at the rear, lasted long enough for a man to get his bearings and start a family. Then he either incorporated this cottage into his new Colonial or relegated it to be a shed or barn.

A cellar, dug and lined with field stones (in a few cases, early brick), was outlined with heavy oak sills, secured at the four corners by mortise and tenon joints. The use of oak beams throughout the house, hand hewn with the indispensible broad-ax, was universal, and ax marks are still there, in any exposed beams, for all to see. (Some

Colonial

houses had a partial cellar under one room only or simply a crawl space.)

Construction of the four room ("two up and two down"), Colonial, continued by the positioning of four corner and two chimney posts, front and rear, mortised into the sill. Connected by cellar, second floor and attic girts, end, front and rear, and corresponding chimney girts through the middle, the frame was topped with rafters. Girts were mortise and tenoned, with rafters double notched into them and dovetailed where they met at the ridge. Every joint was strengthened by a sturdy oak peg, hammered into pre-cut holes.

Inside support consisted of the summer beam, usually set between chimney and end girts, on each level, although occasionally one would span front and rear girts, or be eliminated entirely, in a small room. These were frequently chamfered, removing the sharp edges, largely for safety sake. Chamfering also softened the inside of the end posts, where if flared, they were known as gunstocks for their resemblance to the stock of a gun. Run between summer beams and girts, were the joists to hold up floor boards. These, too, were neatly dovetailed.

There were two methods of framing the roof. No ridgepole was used except where purlins were run horizontal to the peak. Common rafters, joined and pegged at the ridge, are sometimes considered older, although this conclusion is controversial.

This then, is the basic framing of the Colonial and sections of it are generally visible in all of these houses. Indeed, where they do show, they are invariably capitolized on as decorative assets.

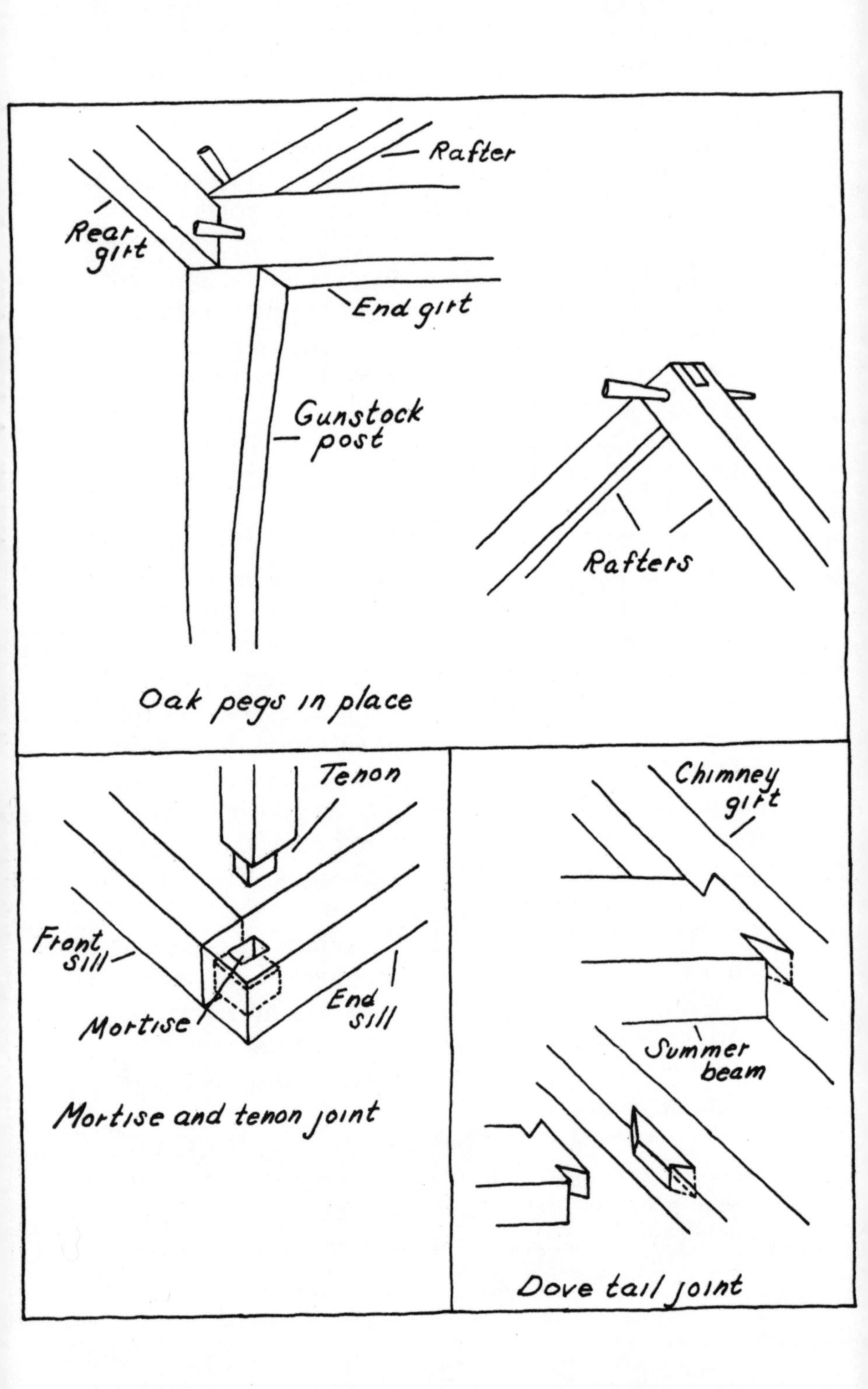

Rafter
Rear girt
End girt
Gunstock post
Rafters
Oak pegs in place
Tenon
Front sill
Mortise
End sill
Mortise and tenon joint
Chimney girt
Summer beam
Dove tail joint

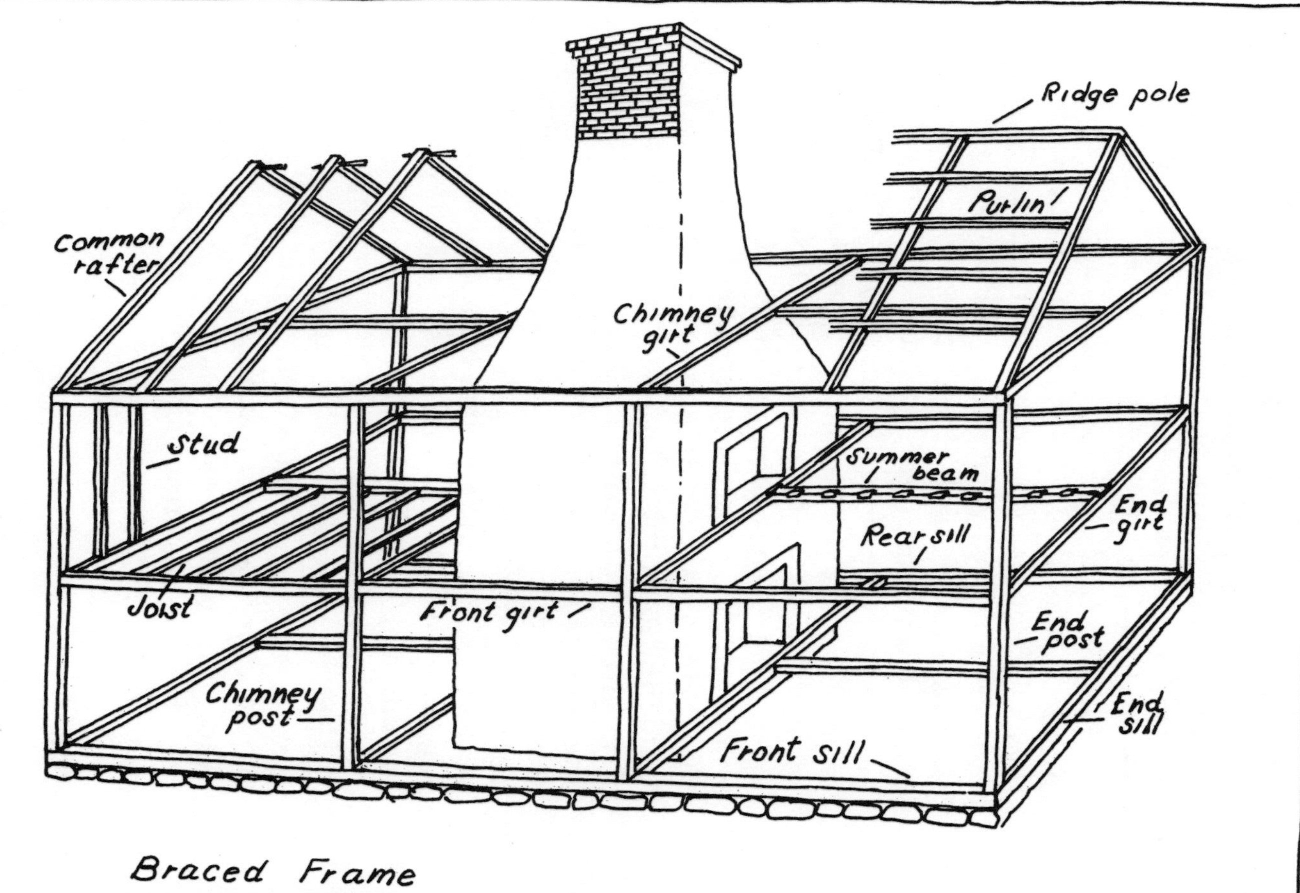

Braced Frame

More often than not, the corner posts, girts, and summer beams project into the rooms, and where these beams are not cased and edged with molding, even the humblest of them show off their original beauty. Casing became standard practice in grand houses by the early seventeen hundreds and in the vernacular by mid-century.

Nestled in the middle of this carefully constructed skeleton was the center chimney, making itself obvious in every room of the house. The earliest chimneys were built of field stone with the great square base filling up to one third of the cellar, depending on the size of the house. Later brick ones were of equal size. Sometimes field stone chimneys were topped with brick above the second floor or from the attic level. A fireplace opened on to every room, with the chimney accomodating a flue for each, terminating above the roof, behind the ridge line in some early houses, straddling it in others. The chimney, snuggled up to its girts, left a small space behind at the back of the house and only a slightly larger space in front, where the stairs crowded up against it, turning corners, narrow and steep. This chimney dominated the house to help ward off winter chills, by retaining some heat when fires were allowed to die down at night.

Floor boards, paneling or sheathing, roof boards and those around the outside of the house, were of pine, some quite wide. While clapboards of oak gave way to those of pine, on front and sides, pine weather boarding covered the back, and the roof was crowned with oak, chestnut or cedar shingles. The simple roof line was occasionally broken by large dormers, more like gables at right angles.

Variations of this house were built from the second half of the seventeenth to the early nineteenth centuries. When evolving from the "one room and loft" cottage, this Colonial was seldom as symetrical as the "two and two" built as originally designed. Hence, an off center chimney or irregularly placed windows suggest that one side might be older than the other.

No two Colonials are exactly alike, because they were built by, and for, different persons for their own use. Modifications of detail are due not only to the individual builder, however, but to geographical location, and change, over time, in technology. Windows with many panes that always distinguish the Colonial, started very small, with tiny panes, when glass was imported from England, expensive and scarce. Later as we began to make glass here, the "six over six" was enlarged to six over nine and still later, to nine over twelve. In the few houses where lights occur over the front door, they often salvaged thrown away panes, known now as bulls-eye glass and jealously collected. Bulls-eye glass was perfectly adequate to let in light, where one did not have to look out.

Some pine clapboards still survive with feathered ends that overlap, but most have been replaced with the later butt end variety. During the mid-seventeen hundreds, it was "derigueur" to graduate the width of the clapboards, starting with narrow ones at the sills, while gradually increasing the width to the eaves.

The front door sometimes opened into an enclosed projecting porch or vestibule, reducing draft in the small entrance hall. Plain frames around doors and windows, inside as well as out,

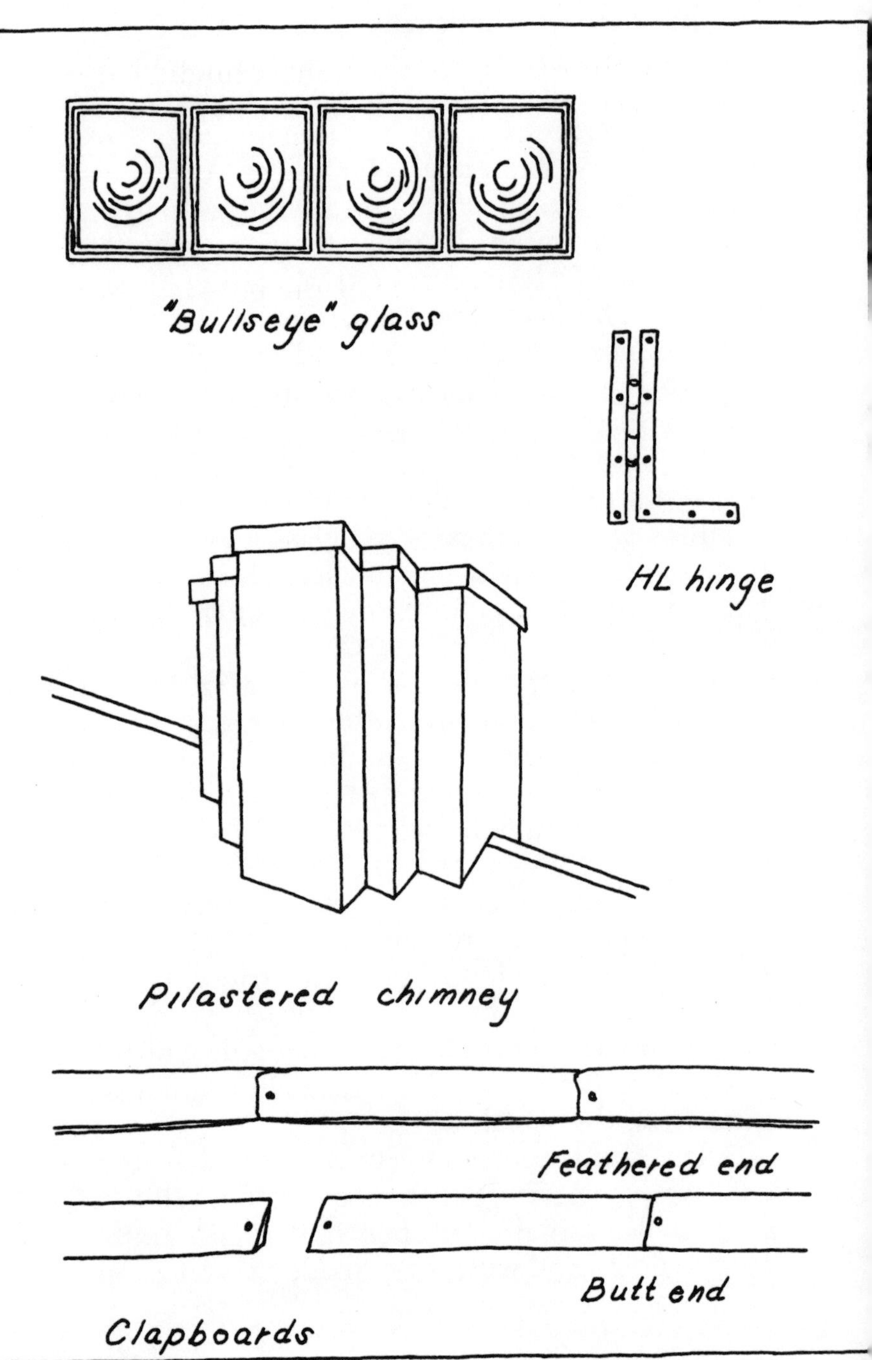
"Bullseye" glass
HL hinge
Pilastered chimney
Feathered end
Butt end
Clapboards

complimented the simplicity of the Colonial. No garnishing here, for reasons of ease in construction, as well as lack of descretionary funds. These houses were essentially all farm dwellings and because they housed large families, pennies counted.

Some original boards used in flooring and sheathing were hand sawn by the pit saw that left slanting ridges on the face of the board. Others were processed by water powered "up and down" saws, in operation prior to seventeen hundred, and are distingushed by up and down marks.

Plastering (with cow or horse hair filler), that appears in the Colonial house, was on split-sheet lath until sawn lath took over in the nineteenth century. Handmade nails and smithy hardware lasted until eighteen hundred and frequently later. HL hinges also were a favorite for one hundred years, from the early seventeen hundreds.

Salt Box

The Salt Box or shed roof feature was a natural outgrowth of the Center Chimney Colonial, when a lean-to was added for more space. As the size of the family grew, so did their house. Shed roofs, always added to the north side of the house, helped to deflect cold winds up and over the house and discouraged the build up of snow, contributing immensely to the comfort of its sheltered family.

At first the shed rafters were dovetailed and pegged into the ends of existing roof supports. This may or may not have produced a straight line from the ridge to their termination at a point

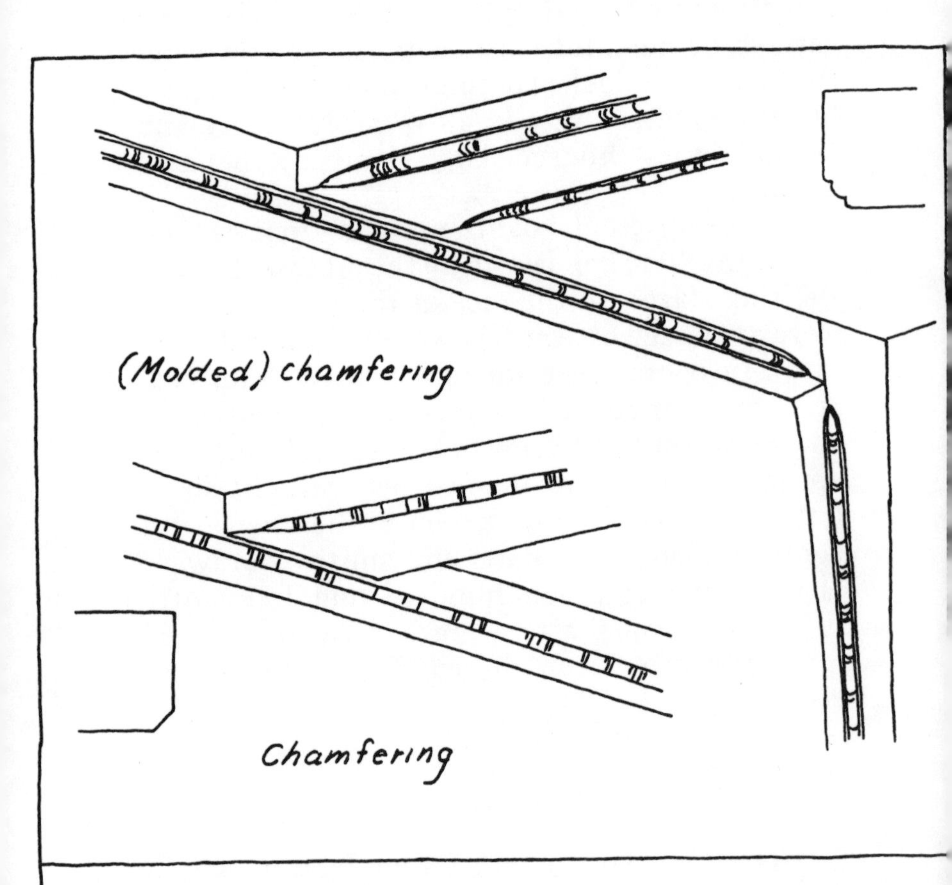
(Molded) chamfering
Chamfering

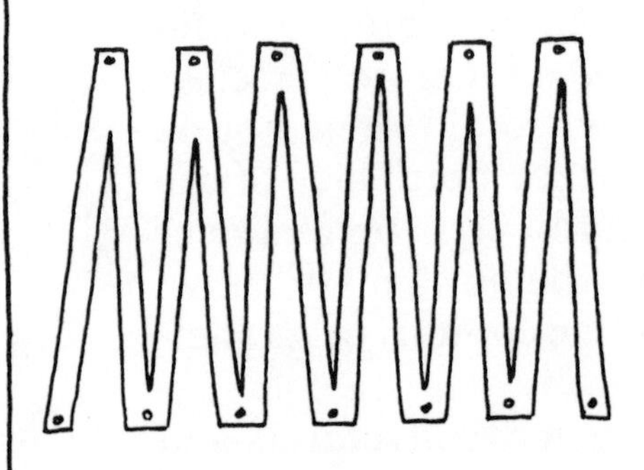
Split sheet lath

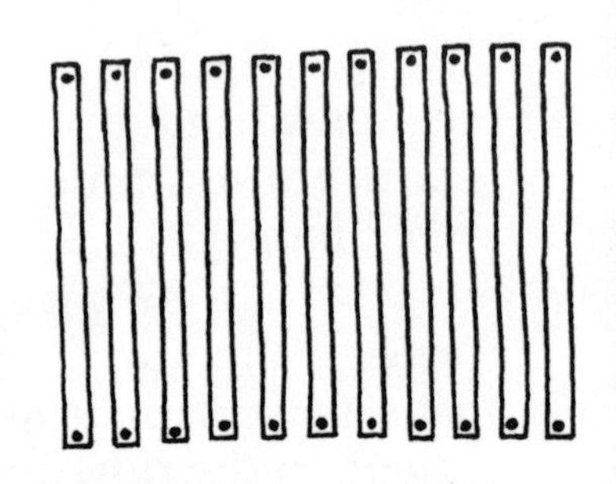
Sawn lath

Salt Box

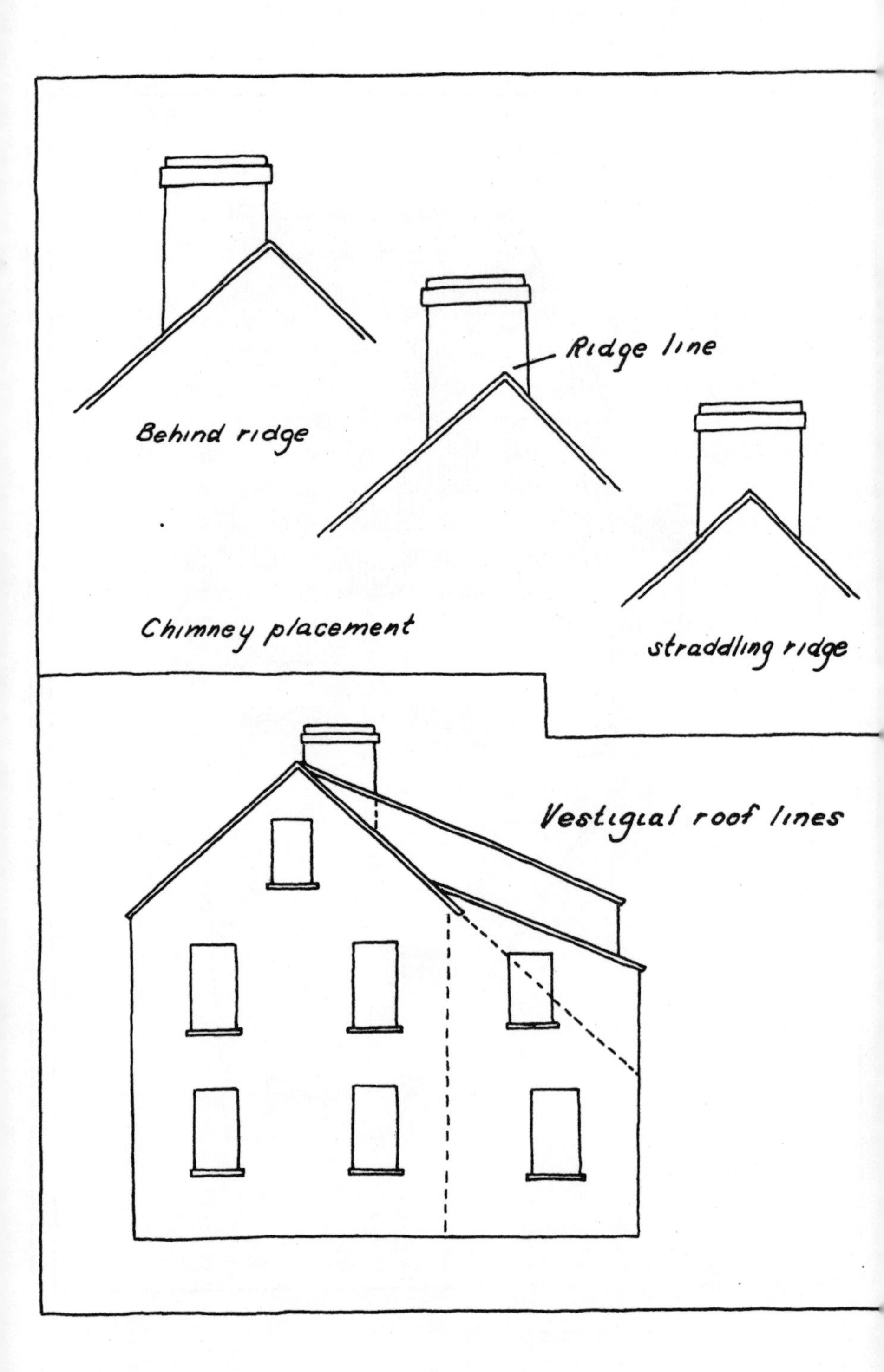
Ridge line
Behind ridge
Chimney placement
straddling ridge
Vestigial roof lines

slightly below the second floor level, or in some cases, lower. A roof with a broken line, however, is a sure sign of an added on back, indicating that the front is somewhat (or much), older than the shed. In later pre-planned versions, the shed rafters were one continual piece from the ridge to an arbitrary rear second floor level, as the pitch could be more easily lifted. Lean-tos varied in width, from one room to two, or two and staircase, culminating in a two story projection at the side of the house.

The resultant new area became a spacious kitchen and smaller "keeping room", with two mini step-down chambers under the eaves upstairs, tandam to the front rooms. They were useful for not much more than storage. A generous walk-in fireplace was then built into the new kitchen, with a second girt juxtaposed beside the old rear sill to carry its weight. And a new flue was run up the chimney at the back, forming a noticeable pilaster on the original stack.

If the shed projected beyond the width of the house, a new pair of stairs led up to a narrow hall behind the chimney, opening on to one or more rooms.

All construction techniques and details of the Salt Box remained the same as its predecessor, as it was nearly as old a style and in any case overlapped, by seventeen hundred, and for many years thereafter.

Both the Center Chimney and Salt Box Colonials yielded, in time, to additional rooms attached to the side or rear, and in both, a part or all of the shed roof has been raised one or more times to make the eaves rooms more functional. Sometimes a vestige of the original roof line has been

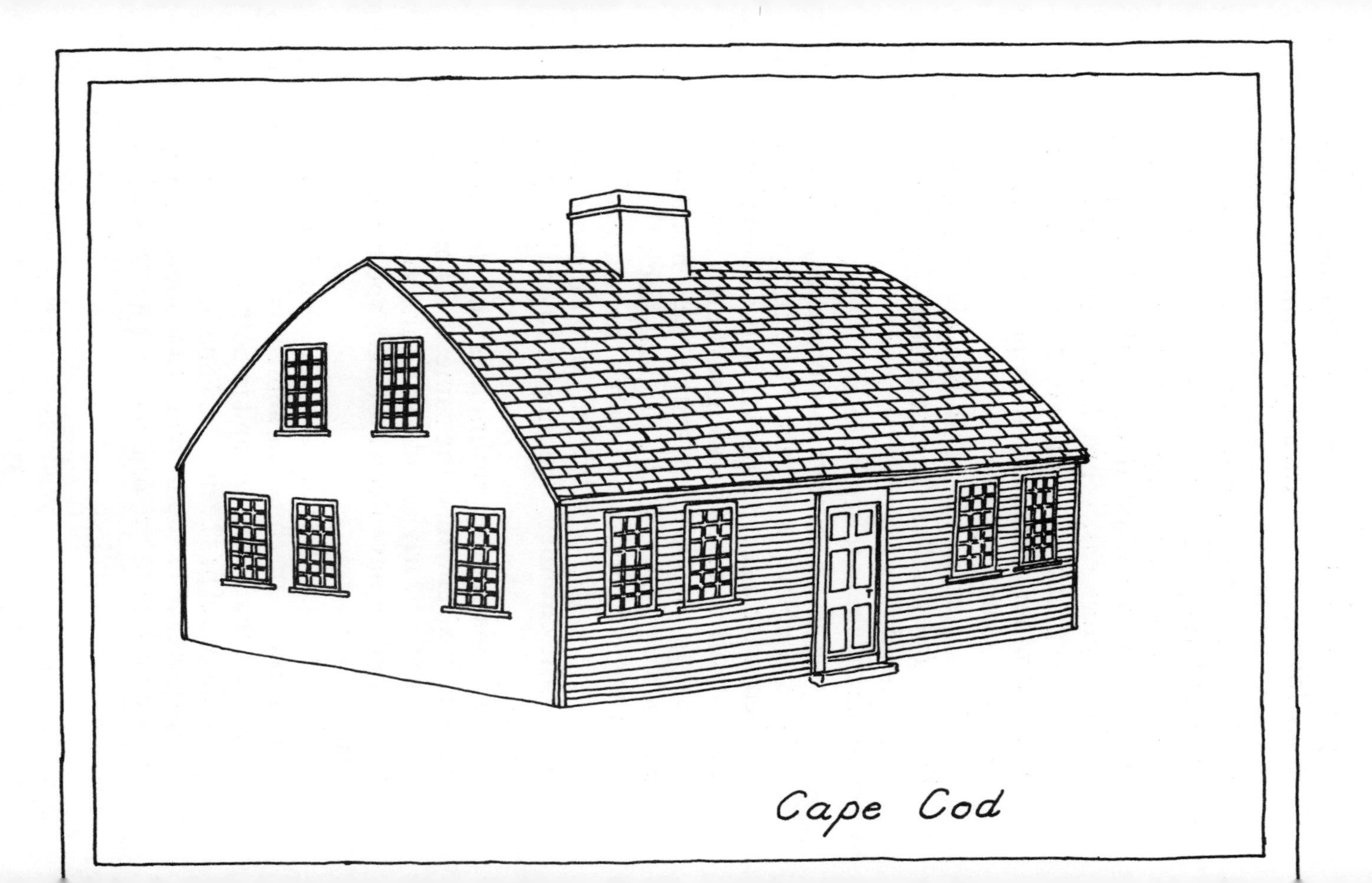

Cape Cod

preserved, but in many cases, new sides are flush and clapboards conceal that dramatic sweep. An attic sleuth may detect the line, on the inside, never-the-less; perhaps find a sawed-off rafter and peg, still extant.

The seventeen and eighteen hundreds brought on a rash of "modern" kitchens (at the rear or side of the house), with their own fireplaces and the inevitable string of sheds ending at the barn. In the seventeenth century, some towns passed a strict law against connecting house and barn, in order that one might have a chance for survival, if the other burned. But by the nineteenth century, even though fire fighting methods had not improved appreciably, the law no longer applied as people relaxed their vigilance.

A spin-off of the Colonial, that combined features of the Salt Box, was a floor plan, two rooms deep upstairs and down. This transitional layout had one or more chambers at the rear upstairs, usually with a second pair of stairs at the back of the house. One chamber might have a fireplace, but others necessitated a second chimney ofset behind the ridge, perhaps from the downstairs keeping room.

Cape Cod

A classic Cape, by any standard, is nifty. Although not initially designed for the customary large family, the Cape did not escape the rambling after thoughts that all dwellings of that era were vulnerable to. Originating on Cape Cod and "The Islands", this unique architectural type, popular from the late sixteen hundreds through the middle of the nineteenth century, is typified by beautiful examples which abound all over New England today.

Again, the heritage, and thus the fundamental design and structural elements, are the same as the Center Chimney Colonial, with certain notable exceptions. The Cape, being four-square, had an additional middle girt at all levels with corresponding summers (where needed), and joists. The center chimney remained in place, straddling the ridge, and was surrounded by four rooms on the first floor, characteristic of the Salt Box. The roof, however, rising from the second floor level at a more rakish pitch accomodated only two rooms upstairs. Essentially, it was a one story house, though the width did manage to allow for a couple of rooms on the second floor with a window or two apiece; not luxurious space, but adequate for bedrooms.

An intriguing anomaly, though perhaps with a purpose, was the half Cape and related three quarter. Unlike the master plan of centered front door and chimney and complimentary windows either side, the half Cape featured the front door and chimney flush with one side, keeping company with the usual two windows opposite, but minus the palancing portion. It was undoubtedly the fervent wish and expectation of its builder to complete the symetry at a later date. Curiously, many of these half houses were enlarged, not conforming to the original intention, but by a thoroughly delightful and unconventional el or wing.

Other variations included the unbalanced floor plan, rooms of unequal size, and the bow roof. This roof respected the clean no dormer rules, but had a slight lift between eaves and ridge, reminiscent of the bottom of a ship. A daring analogy, this may have been true since many of the builders were sea-going men. Also the prevalent seeming lack of inclination to greatly enlarge the Cape, may have been due to the tragic death at sea of many a young father, leaving behind a smaller than usual family to inhabit his jewel.

A final interesting aberation is the fact that a few els appear to pre-date the house proper and may have been, like the earliest "one room and loft" Colonials, bed and board, while the owner gathered courage and cash to tackle the larger effort.

Early Georgian 1720-1760

Georgian architecture introduced irrevocable competition for the beloved Center Chimney Colonial and symbolized a growing wealth, or at the very least, some extra pounds tucked under the mattress. Those not affluent enough to build new, simply retrieved the cache and remodeled their Colonial. But such plain fare no longer satisfied a more sophisticated appetite.

The eight room Georgian thrust two chimneys onto the horizon, added two full sized bedrooms on the upstairs back (which must have been welcome by crowded families), and achieved a new touch of dignity. The only change from Colonial construction was the adaptation of the frame to two sets of chimney girts instead of one. Each chimney flanked the front door, centered on the rear wall of the front rooms, in wooden houses, and single or double ones at the gables, in the brick edition. Left over interior

Early Georgian

space that once housed the center chimney, became a spacious full length hall.

Even though the gentle modesty of the Colonial was still popular among those of limited income, the new formality in a less yielding Georgian design appealed to the rich. The facade of the house was perfectly balanced; front door in the middle, windows equally spaced either side and upstairs windows directly above those below. Even the sides carried through the pattern, with the whole set up on a high foundation.

With the time at hand for a few restrained decorative details, the paneled front door became larger with lights or fan above, enriched by a pediment and fluted or plain pilasters. High in the eaves, grazing the upper story windows, ran a cornice, often with dentils. Many paned windows, were set off by an architrave with the occassional touch of narrow flat cornices. A pitched or gambrel roof (sometimes high hip with balastrade around the ridge), might sprout small symetrically spaced dormers with triangular pediments.

Inside, the stair case swept up from one side of the stately hall, with paneling following to balance the mahogany hand rail. Paneling was also correct for many of the rooms, on one or more walls, with the rest wallpapered or painted. Pine gave way to hardwood floors; doors and windows gained a modest architrave, and fireplaces were adorned with tile and marble.

For those who could not afford the whole treatment, yet wanted the flavor, the vernacular managed the two chimneys, the floor plan and perhaps a paneled door with fan light. The thriftiest stayed put in their old Colonials, removed

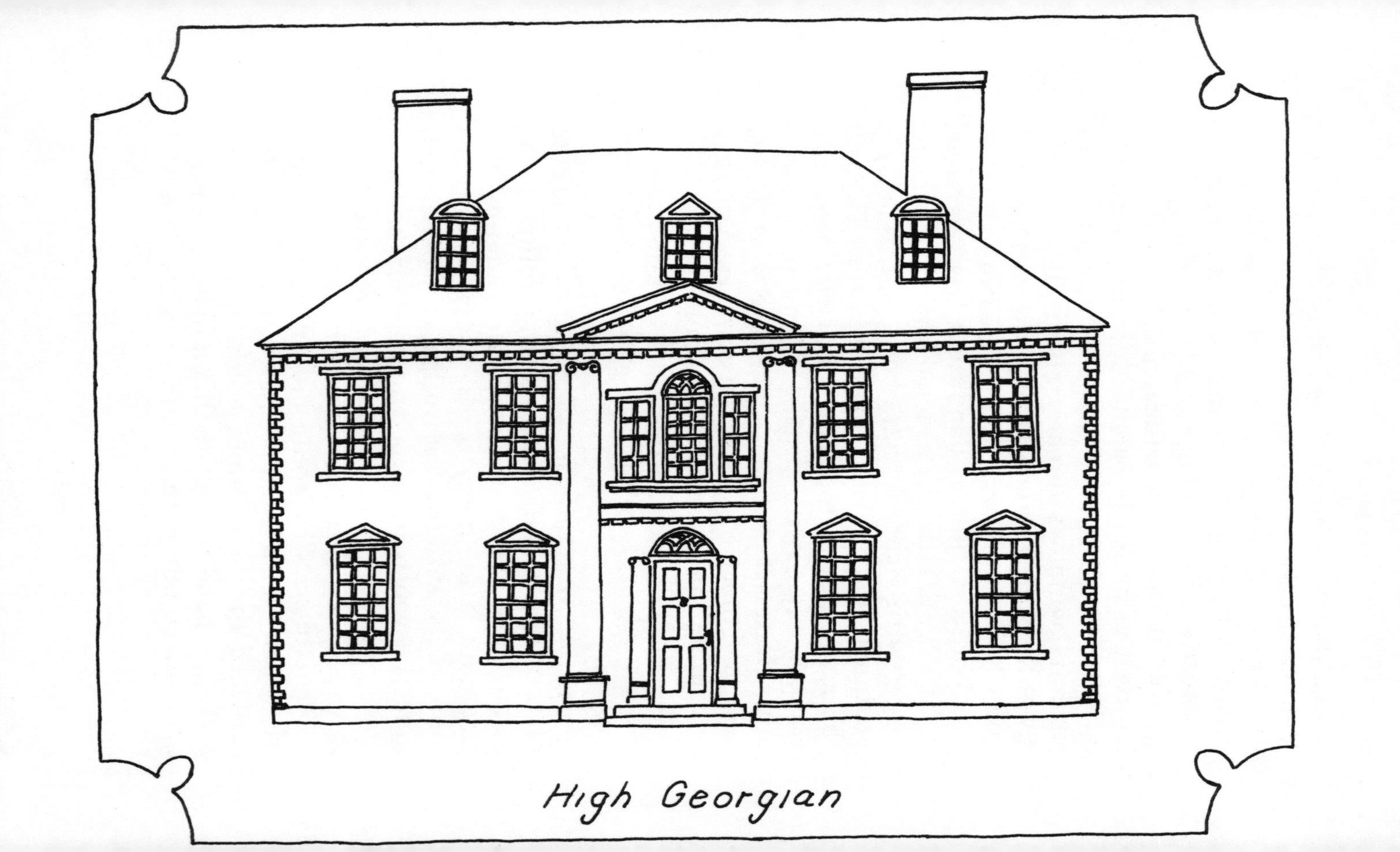

High Georgian

the center chimney completely to form the entrance hall and ran up the two chimneys, one on either side with new smaller fireplaces. In so doing, the kitchen was moved out to an el or wing, in keeping with the Georgian floor plan, and the stage was set for the added-on look. This peculiar and difficult kind of remodeling, to substitute one chimney with two, was also carried on through the time span of the next two architectural styles.

High Georgian 1760 - 1780

As if the taste of elegance was not enough, a full meal soon had to be provided in the souped-up grandeur of the High Georgian. The basic layout stayed the same (plus emergence of a sometime third story), but the ornamentation increased, several fold. This tentative third story gave birth to the "Captain's house", famous in sea coast towns.

An imposing Palladian window above the front door, flanked by columns or pilasters the full height of the front, set the pace. The front central door pediment grew bolder, and sometimes was faced with a double scroll. Pilasters immediately beside the door were flutted or replaced by columns in relief, corners of the house acquired quoins or more pilasters and windows were capped with a more obvious cornice. The roof, in swell society, might have an imposing pavilion with a pedimented gable projecting above the Palladian window, and resting on the master pilasters. Dormers, with or without the pavilion, had alternating triangular and arched caps. Cupolas too, appeared atop the roof.

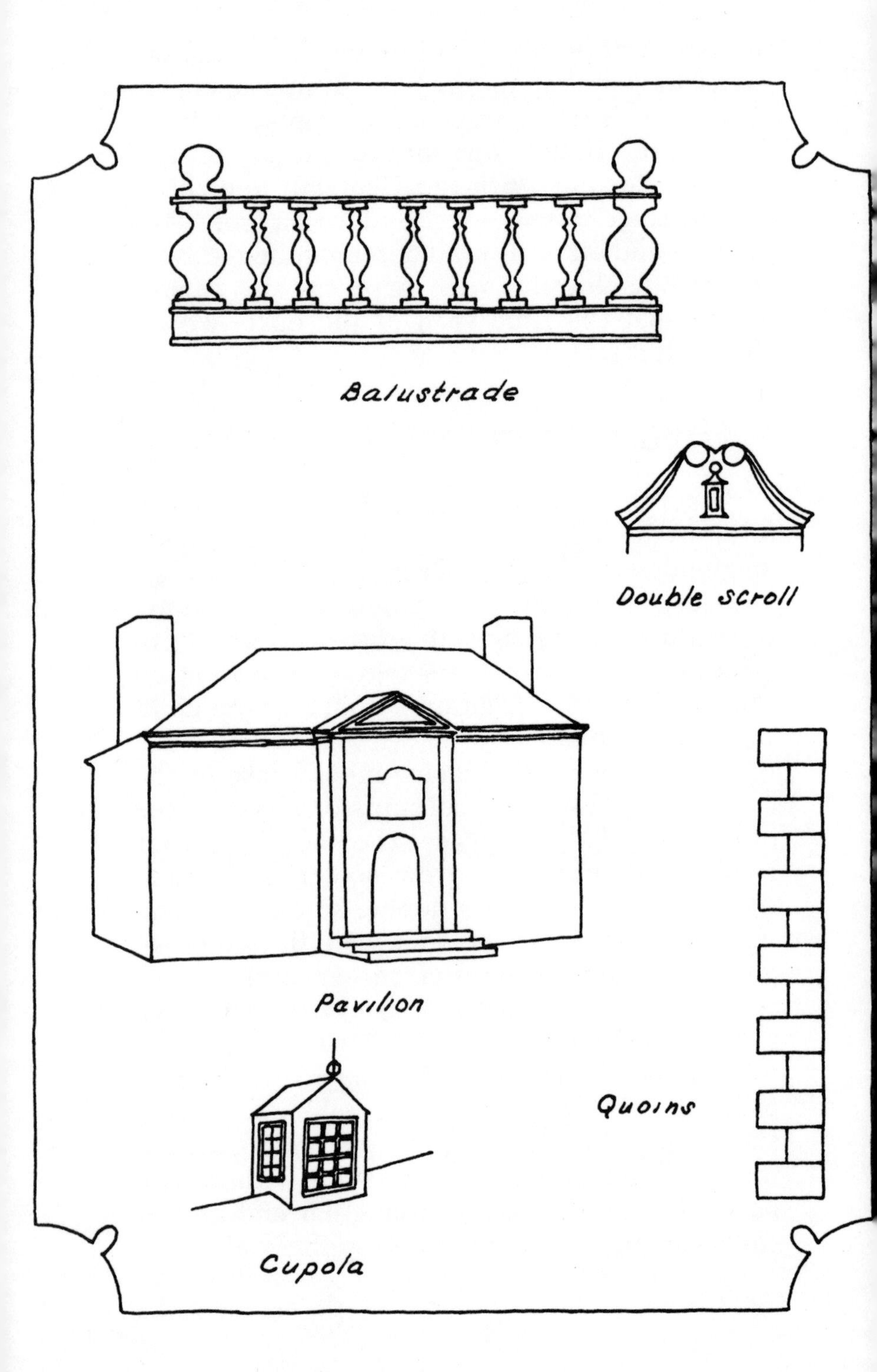
Balustrade
Double scroll
Pavilion
Quoins
Cupola

Federal

Paladian window

Door lights

Not to be outdone by the exterior, the inside of the house frosted the cake of this splendid feast. Staircases had a variety of elaborately turned balusters with the handrail flairing out at the bottom to envelope the equally ornate newel. Pilasters guarded pedimented door ways and paneling above the fireplace, and cornices surrounded the room at the ceiling. Every room was heavy with regal decoration.

The vernacular High Georgian was often the Early Georgian dressed up; in many instances, well disguised.

Federal 1780 - 1820

With appetites satiated, the orgy of embellishments diminished. Federal refinement of the Georgian additives tidied the clutter and gave selected details more class, presaging the reactionary severity to come.

Two and a half and three storys (brick or wood), remained popular, with the roof more often flat hipped, hidden by a balastrade at the eaves line. The pediment over the front door kept its pilasters or classical columns, but became a flat entablature, frequently projected in the form of a sheltering portico. The lacy eliptical fan over the door spanned new side lights and window panes were enlarged. Window frames also were simpler and the corners of the house gradually lost their quoins and pilasters as the wall fabric changed to flush boarding or painted brick and stucco. The quieter taste of the outside was reflected within in the more restrained and airy freedom.

Curves began to creep into the interior of the

Dado
Reeding
Molding
Molded plaster
Sheaf of wheat
Newel post and baluster

house and an irregular floor plan did not bring on the wrath of God, as it seemed was earlier feared. The rooms blossomed with more descriminating, less flamboyant details. Wall paneling was subdued to a flat dado topped by simple molding, mantles and overmantles smoothed out with architrave bands and the merest hint of delicate reeding. And new, also, were the personal touches, in fancier homes, of molded plaster flowers, shocks of wheat and dainty garlands, painted white to match the rest of the woodwork.

The more flexible floor plan included a service wing, with additional chimneys and such daring features as oval rooms and alcoves. Experiments so bold leave us a legacy of simple elegance and livibility in these oversized houses that we can no longer afford to maintain or enjoy in the gracious tradition once possible.

In the vernacular, the Federal was a farmhouse with paneled front door topped by a small fanlight and high entablature.

Greek Revival 1820 - 1860

As if the stern Mother-in-law had come to visit, the party was over and it was time to end the frivolity. Many "house builders" in this sobering period prefered to call themselves, more professionally, architects. The design of correct proportions and linear balance of the Greek Revival was aided by American architectural publications and the building of a house continued to develope as a serious profession first introduced by a few elite architects who used British plans of the Federal period.

Greek
Revival

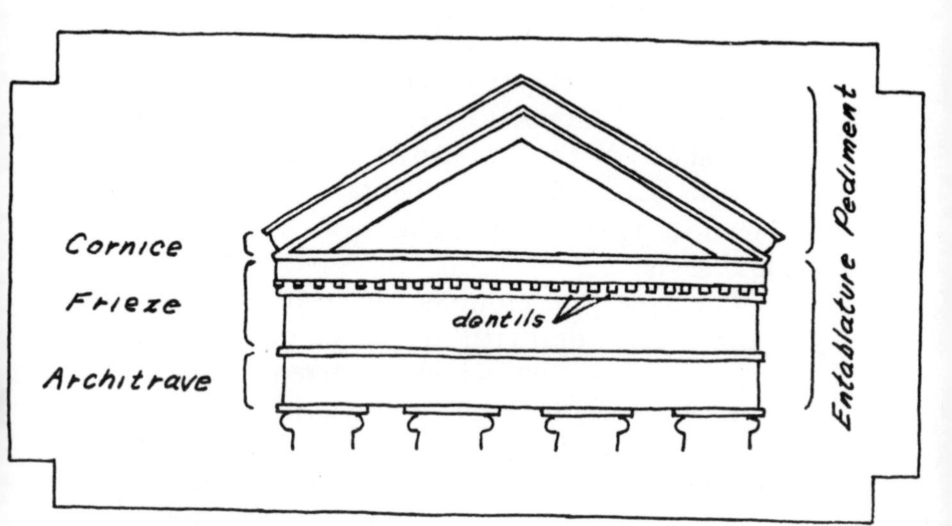
Cornice
Frieze
Architrave
dentils
Entablature
Pediment

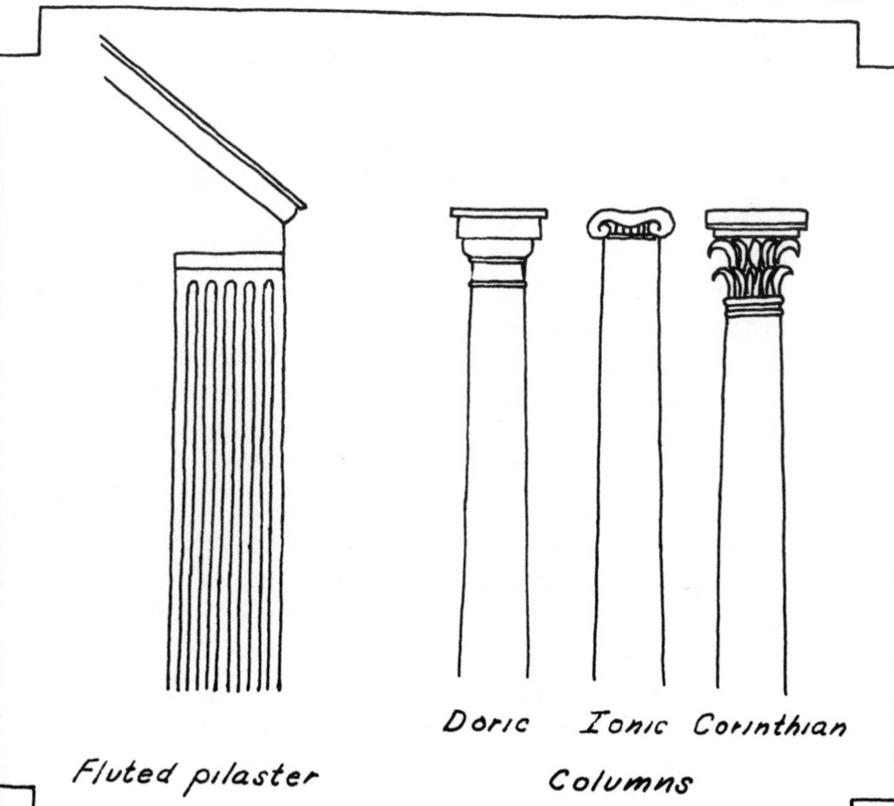
Fluted pilaster
Doric
Ionic
Corinthian
Columns

In some ways, Greek Revival was a partial return to sanity, and although the large Georgians were not generally a concern of the average home owner, Greek Revival even deplored the waste space of the overgrown farm house. The versatile Greek Revival, adaptable to mansions as well as smaller dwellings was a totally national style.

The switch of emphasis from front expanse to gable end, dramatized the change brought about by Greek Revival design, even if this was not true in every case. The fussiness of the Roman Georgian was dropped for heavy and forthright Greek asceticism. Suddenly thrift and clean lines became imperative, so the narrow gable end facade necessitated a small side hall, placing the front door to one side, balanced by two windows. Second floor windows of the same size were evenly spaced under the gable which was sometimes rather forbidding. This gable was either flat, or projected forward to rest on a row of Ionic columns forming a portico. A flat pedimented gable sat on corner pilasters or simpler broad corner boards. In larger and more elegant houses, while the gable end faced the street, the front door stayed in its traditional place and the floor plan remained Georgian. Chimneys were minimized and some roofs flattened with high architrave and wide plain cornice. The simplicity, if on somewhat robust lines, was carried through inside as well, with plain paneling and the use of Greek fretting and rosettes around doors and windows.

A few smaller one and a half story houses sneaked in a row of tiny second floor windows under the eaves line. These houses also augmented the limited number of rooms inside by an off-set el at the rear. Thus was begun, what was called the first All-American style of architecture and its application ranged from the humblest dwelling house to the mightiest of public monuments.

Once again, the vernacular appeared in a rather strange form. It might have been an old Cape, newly raised half a story to incorporate a row of half windows and a pedimented gable, or a Colonial with pasted on gable and an Ionic columned front porch. Some farm houses had Greek detail around the door, two thirds length sidelights and a flat entablature with distinct architrave, freize, and cornice divisions.

Gothic Revival 1835 - 1880

With hysteria for maintaining the "status quo", at an end, change became infectious. Architectural innovations introduced by Gothic Revival were not simply token or superficial, but had real meaning in terms of the pocketbook as well as for family comfort.

Although cuteness of style and self-conscious trim tend to turn off some people, an easy going floor plan supported the rational argument for designing a house from the inside out. Function became the keynote that coincided with the unveiling of a cheaper balloon frame and new fangled stoves.

The circular saw which could process as much wood in a day as it took the up and down saw a week to do, reduced cost and turned out lumber of less cumbersome dimensions. Nails, that here-to-fore had to be made all by hand, were

Gothic Revival

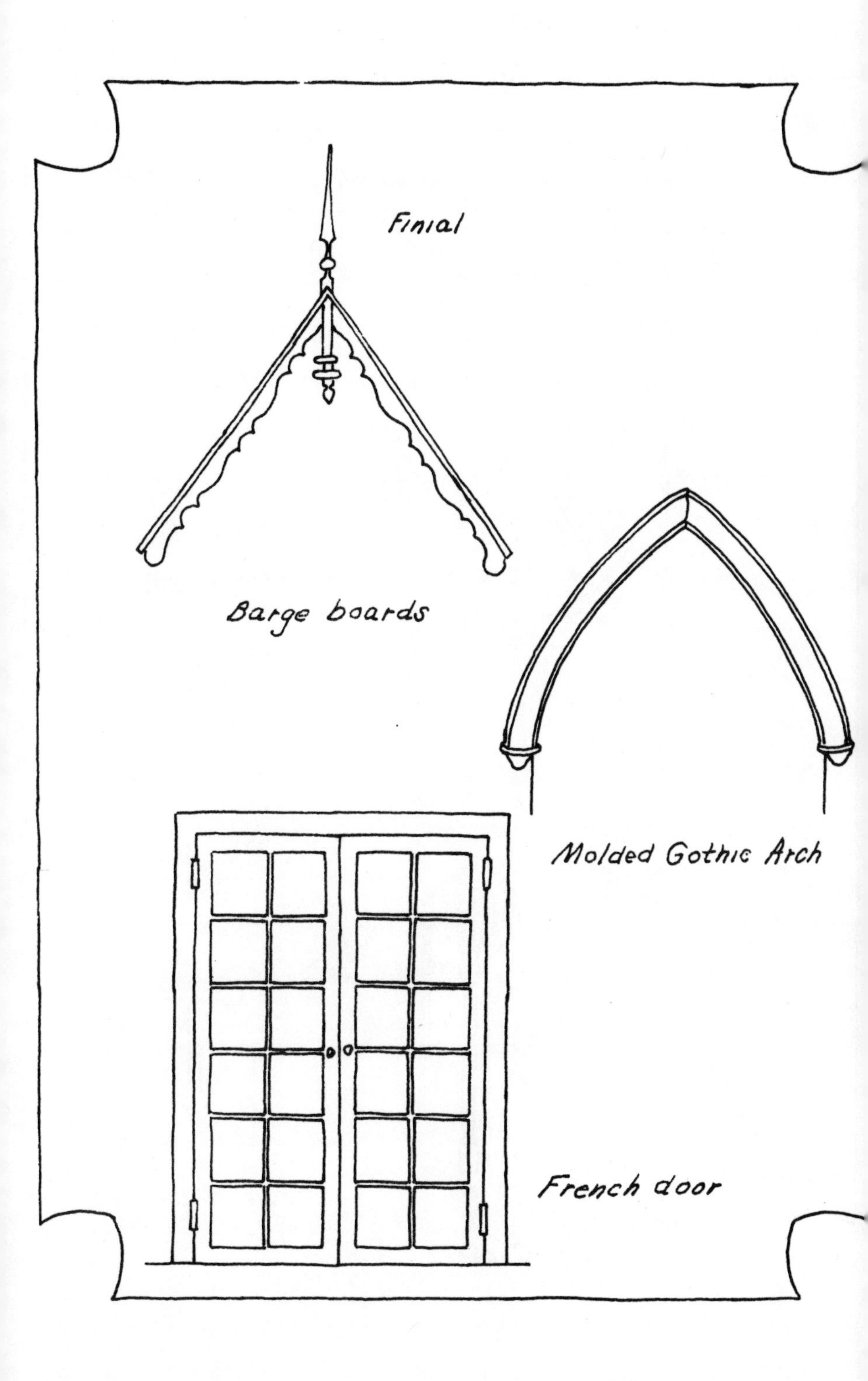
Finial
Barge boards
Molded Gothic Arch
French door

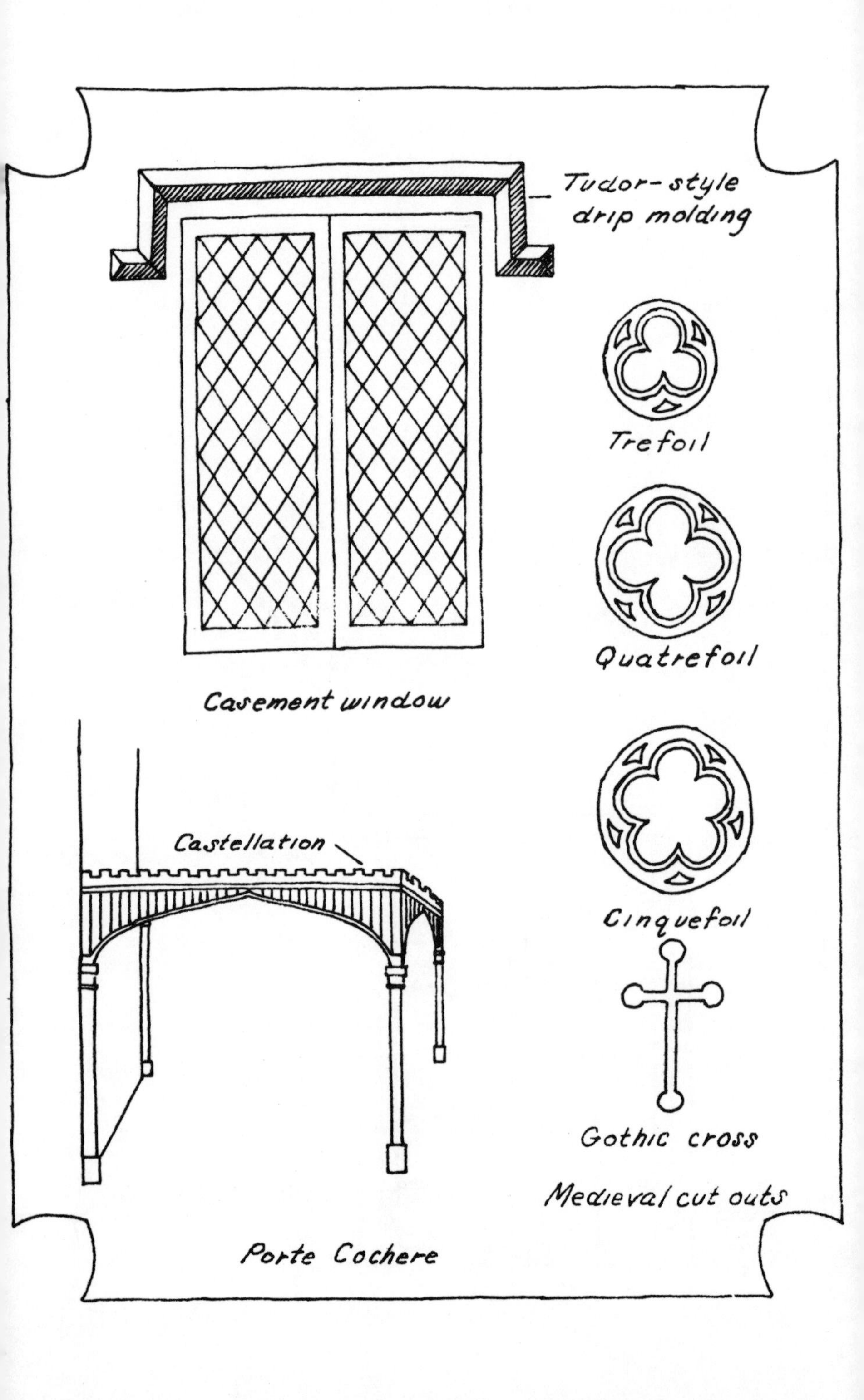
Tudor-style drip molding
Casement window
Trefoil
Quatrefoil
Cinquefoil
Gothic cross
Medieval cut outs
Castellation
Porte Cochere

being mass-produced by machine. The initial Gothic Revival houses, built around braced-frame construction, yielded rapidly to those utilizing the new balloon frame. This method of framing a house used smaller members, nailed together, obviating the need for mortise and tenon joints and dove-tailing. Construction was faster, hence less expensive, and much lighter. Chimneys slimmed down to vent stoves only, as fireplaces were phased out, and the floor plan adjusted to additions in any direction with a multiplicity of gabled windows for bright cheery bedrooms. The less formal layout could suffer growth while achieving balance with imbalance.

Exterior features were usually dominated by sharp pointed gables, topped by a slender finial balancing a similar drop hanging down inside the peak that joined scrolled barge boards cut by the band saw. A molded Gothic arch set off the front door, and windows (some with leaded glass casements), were not infrequently topped with a Tudor-style drip molding.

Three to five sided bays opened up reading nooks inside, while porches and the occassional castellated porte cochere provided plenty of excuse for trim. This decorative woodwork displayed medieval cut outs; Gothic crosses, trefoils, quatrefoils, and other tracery. Subdued behind the bulges and garniture, vertical board and batten faced the house.

Inside, window seats conformed to the bay windows, hardwood floors gleamed, double parlors were connected by sliding doors, while French doors opened onto trellised porches.

"Carpenter Gothic" was a term applied to the simpler and most naive of these houses in

Italianate

which no attempt was made to adhere to traditional designs. The more fanciful the "gingerbread" that was cut by the scroll saw, the better, and features such as turrets or spires were added with no regard for any original function or logical placement.

Gothic Revival vernacular slipped in a couple of sharp gables, a Gothic arch over the front door or a small touch of medieval tracery stuck on to porch supports and barge board trim in the gables.

Italianate 1845 - 1885

Vying for equal status, the so-called "Italian Villa", joined the spreading architectural eclecticism of the middle eighteen hundreds. Angular and asymetrical, it gave prominance to a tall square tower to emphasize the vertical of its other elements.

The Italianate (or "Bracketed" style), presented a strong bulky appearance which tended to be over embellished, even though these houses were frequently painted in pastel colors to simulate a mediterranean air. The front door was cloistered by a porch with its flat top supported on square columns, which in turn held up semicircular arches. Single or clustered windows, with arched tops, sometimes sat behind heavily bracketed balconies and a whimsical oriel window might be set in a first or second floor niche. Ornate brackets also bolstered exagerated eaves that followed the flat or low pitched roof. Early brackets appear singly and were cut in simple curves. Later ones, of more elaborate turning, were paired. Horizontal decorative bands, to-

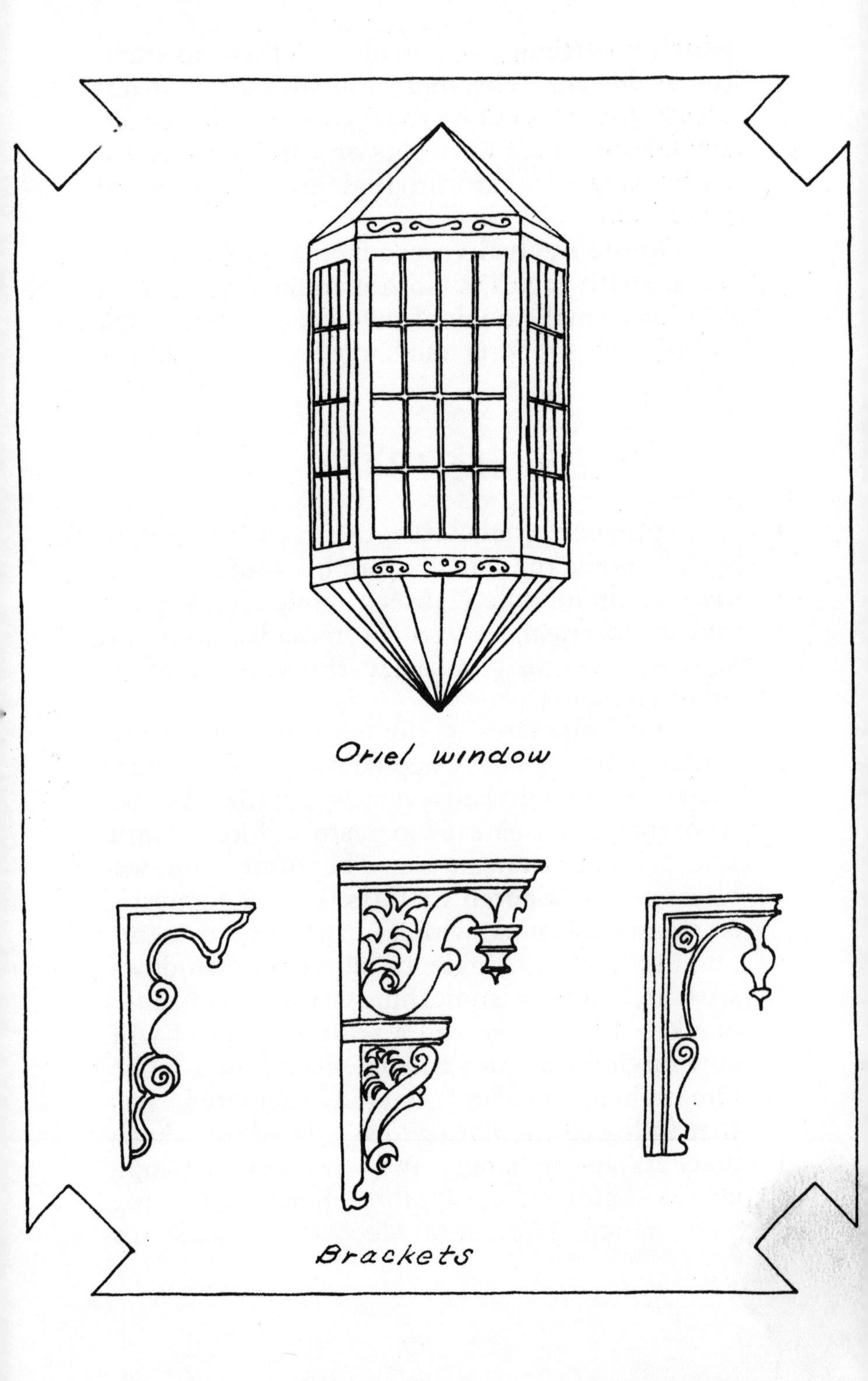
Oriel window
Brackets

gether with a low lying piazza or wing, sometimes broke the vertical thrust.

Some Italianates were boxy and rather formal on the inside (having been built with the heavy Colonial frame), but high ceilings, full length windows downstairs and easily opened casements upstairs, softened its harshness. Even so, the inside tended to stuffiness with ponderous moldings, dark woodwork and marble.

Mansard 1855 - 1885

Anxious also, to advance change, the Mansard style was named for a distinctive roof with its steep slope broken half way for a gentler drift toward the ridge. Usually this roof went around all four sides of the house and incorporated in it an amazing variety of dormers. The simpler windows had round or pointed caps or were gabled. Others were clustered, sometimes projecting to form a bay or even an independent turret the full height of the house. The Mansard roof broadened attic headroom to encompass a fully functioning second or third floor and often covered with slate might have an encircling balastrade. Some fancy Mansards were symetrical with central pavilions and flanking wings, while others were laid out in axial planning.

Many larger houses of this style were rather lugubrious for all their attempts at lightness with oversized windows, a touch of colored glass and lofty ceilings.

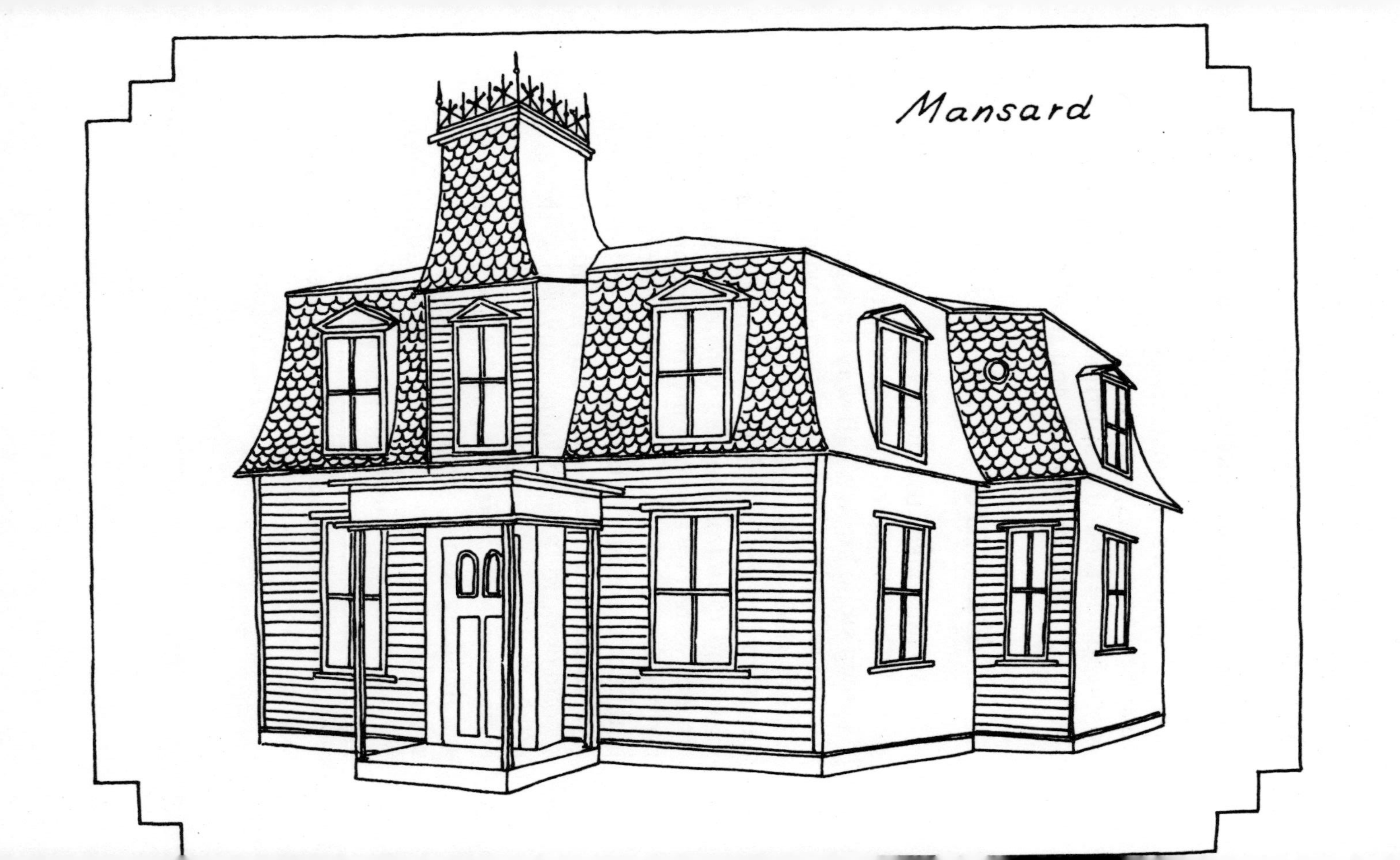
Mansard

Stick Style

Shingle Style

Queen Anne 1875 - 1900

The final group of pretenders to the throne were presented together under the name of Queen Anne, or more commonly, "Victorian". The Queen Anne designation embraced several related styles, as well as a specific experimental philosophy, characterized by architectural inventiveness, a delight in the decorative use of materials, including colored glass, and the resurgence of working fireplaces.

The first of these was the Stick style, popular in the late eighteen sixty's, which featured the use of geometric designs composed of wooden members resembling framing but not functional as such. Stick style houses were prone to vertical masses and steep roofs with bisecting gables, irregular porches and bays of one or more storys. The gable peaks and dormers were faced with a cobweb of sticks in a geometric pattern and porch posts were joined under the eaves by similar groupings of sticks. Built in wood or brick, these houses had irregular floor plans and high ceilinged spacious rooms.

In the eighteen eighty's the Shingle style debuted with the entire outer surface of the house covered with shingles, frequently cut in diamond or scalloped shapes. Often large and somewhat rambling, these dwellings had rounded turrets, porches, palladian windows, any style roof with dormers and sometimes gaudy pilastered chimneys. This same massing of structural shapes was also constructed in brick or stone. The informal interior rooms sometimes had small nooks with leaded casement or oriel windows and big roomlike hallways with fireplaces.

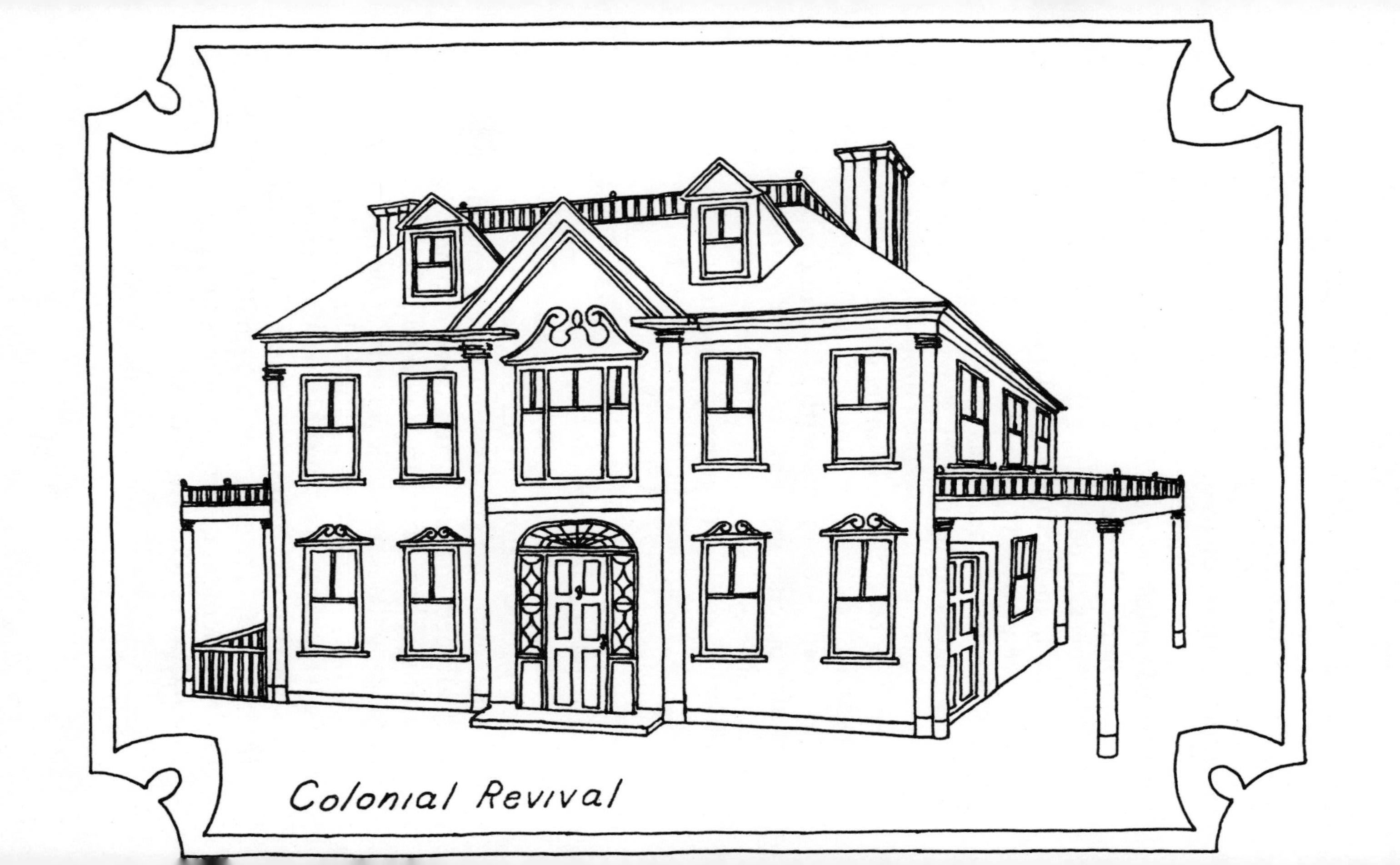
Colonial Revival

A third style was called Colonial Revival and also thrived in the eighteen eighty's. In this era the architects dipped back into eighteenth century Georgian for many of its features while at the same time "improving" on them with irregular additions and details of intricate millwork. Porticos or port cocheres were simply columned and braced with fancifully scrolled brackets. Hip roofed or gambreled, a classical layout was often altered by side porches and one or two story bays, an overhanging second story, or rounded, pillared front portico and an imaginative recessed or bowed out Palladian window. Commodious rooms flowed into one another and again fireplaces and elaborate staircases dominated enormous hallways.

Urban adaptation of the later styles; Federal, Italianate, Mansard, Greek Revival and Queen Anne resulted in row houses. Differing from "houses in a row", row houses were all attached together in the space saving mode. Although built for single family living, they were not all privately owned as some were investment property for rental income.

Not suffering from being hitched to each other, the houses still maintained their unmistakably singular features on the front facade and in most instances a street would be lined with one period. But because the lots were owned by different people, a disparate style could intervene. Also the inevitable fire might have opened up a space to be filled with the "latest" fashion. Many of these places have been altered today for apartment living.

This outline of basic discriptions will answer easy questions, but for the sticky detail that can make or break a specific date, the bibliography at the end of this book contains more scholarly work.

IV WHERE TO DIG FOR FACTS

Have you been able to place your house within a general time frame? Then you are now ready and undoubtedly eager to nail that date down exactly. Fortunately, most of the sources which you need are publically available and waiting.

Deeds and Plans

A deed is a legal document which records the sale, and in some cases, the mortgage of a house and land. Every time any property is sold, a deed is written up and duly recorded in a County Court House. A copy of the deed is reproduced in a Deed Book and referenced in an Index Book. Researching deeds is not difficult once you locate your County Court House and find your way to the Land Registry department and correct sequence of volumes. Your greatest distress will be in the deciphering of crabbed script in some early deeds and understanding the legal terminology of the time. The glossary at the back of this guide will help you with unfamiliar words and phrases.

But, first check in your own files. When a house is sold, a title search is conducted by your lawyer's firm or Title Insurance company. In order to assure that the title is free and clear of all incumbrances, the search will cover several previous owners and from fifty to one hundred years. If your house is less than one hundred years old or has not changed hands many times, you may already have all of the available deed information. Sometimes a house is kept in one family since it was built and the only Deed of Sale will be the one

Know all men be sore whome this Deed of Sale Shall Come Greeting yt I Joseph Eaton of Concord in the County of Middlesex in New England Clerk For and in Consideration of ye Summ of Seventy and two pounds in money

. . . .

Also another piece of Land Containing fourteen acres more or less bounded on the East Side with John Adams & on the South East Side with Frances Brown Land & on the North West Side with Ebenezer Hutten Land wch being added to ye forest tract of Land makes ye whole one hundred and fifty acres more or less besides Seven acres

. . . .

To Have & To Hold The above granted premisses with all the trees Timber like Trees wood and underwood grass and herbage Standing lying or growing upon the Same or yt ever Shall grow upon it having ye fences Cultures Improvements therof fences Water & courses of Water there unto belonging & whatsoever Else thereunto belonging to him ye Joseph Eaton of Concord

Deed sample

that was written up when you took possession. Mrs. Bartlett's house was a perfect example of this phenomenon.

Mrs. Bartlett died at the age of ninety-six, with her only brother preceeding her to the grave. Her house was over one hundred years old and had always been in the family, as had the land before the house was built. No one knew the exact age of the house, or even where the land boundaries were, since the acreage had never been surveyed and a deed was never found. An aged cousin was legally responsible for locating all the scattered heirs; an impossible task even for one able to cope with it. Eventually, a neighbor took over the yearly tax payments and within the requisite number of years, gained tax title and sold the house. It was then that the first deed was composed.

In the case of other older houses, however, the title search may have only tickled the surface of potentially accessible deed material. Now you are on your own. Try not to be alarmed at the buzz of activity that usually is swarming around deed index books. Lawyers and research assistants handle these heavy tomes rapidly and with great assurance, but some are willing to help you with what may appear to you to be a foolish question. Soon, however, you will gain confidence and become as proficient as they.

The Registry of Deeds has records dating back to the 1600's. In doing research, there are two possible methods of attack. Either work backward in time, starting with the people from whom you bought your house, or if you know the name of an early owner, as in the case of some historic houses, work forward and backward

GRANTOR → GRANTEE

Sells Buys

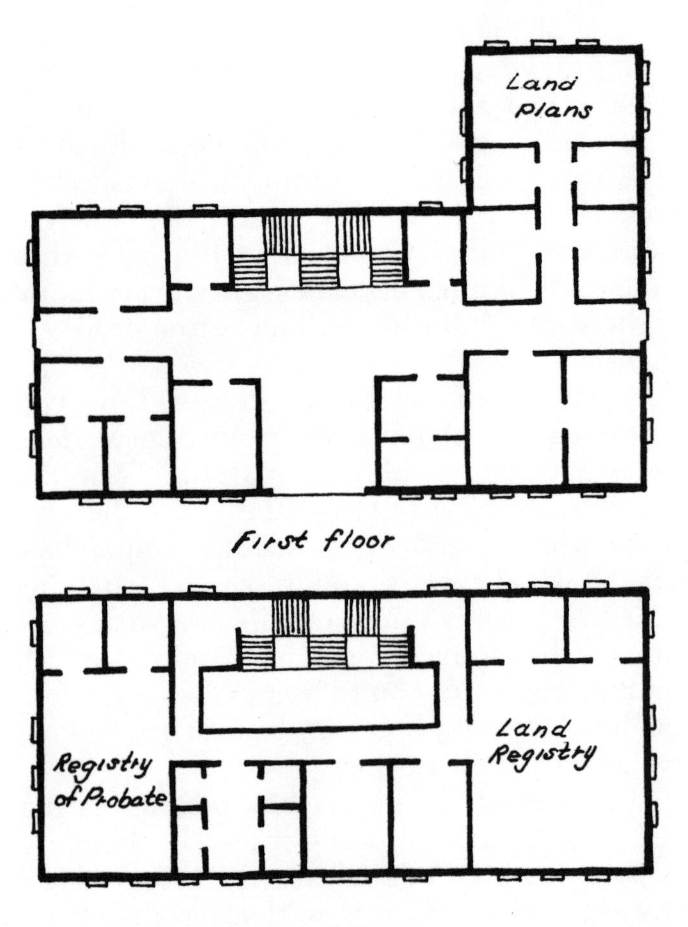

Example of a County Court House

from his name. In the latter situation, you may have difficulty with the date and duplication of names.

An important first step, is mastering the difference between Grantor and Grantee, and remembering that difference! The Grantor sells the property to the Grantee. There will be a double set of indexes, cross-referenced. One set lists the Grantor's name, alphabetized, and the Grantee to whom he sells. The other set is the opposite, with the Grantee's name in order and the person from whom he bought. Unfortunately, the older the date, the less apt the two are to mirror each other. The double set is handy, however, because information missing in one may be found in the other.

If you bought your house from John Wood, look him up in the Grantee index for the year you think he may have bought the house, and you should find the person's name who sold to him. If there is more than one John Wood, you will have to note them all. The date of recording in the Deed Book as well as the book number and page, will send you off to locate the volume that you require. Each deed will in turn state the name of the previous Grantor, enabling you to return to the index to look up his name and continue the process. As you gain expertise, make a list of grantors and grantees at one sitting to save steps. Usually this simple method of research will bring you back to the origin and approximate date of your house or purchaser of the land on which it sits. Sometimes, though, it is not so easy. A chain of deeds can be broken by a will, a series of wills, or the house held in an estate for a long period of time. This is explained under the section on **wills.**

The number of deeds that you find may not necessarily equal the number of owners of your house. Many people in colonial days added to, and subtracted from their estates as their fortunes fluctuated and needs changed, necessitating a deed for each transaction. Some also mortgaged property for a year or more. They may have needed the cash for business or to add a room to the house.

Read each deed carefully, several times, if necessary. Be sure that it refers to your house and land. As the deeds get older, they are less specific about the number and type of buildings and the exact boundaries and location of the land. A deed often repeats the wording of an earlier deed for the same land. This is helpful in making sure that it is the right one, but it can reiterate a mistake. A man selling his land in parcels of several acres, may use the same wording of the first deed that contains his dwelling house and indicate a house where there is none on each of the subsequent ones. Wording such as, "To Have And To Hold, with all the timber, trees, fences, wood, underwood, herbiage and *messuage* (dwelling house) . . ." is so common that it is often automatically lifted out whole from one deed and dropped into any number of succeeding ones.

Deeds can be full of information, or quite stark. Usually they will spell out the number of acres and a description of the boundaries. Old deeds often give the Grantor's and Grantee's occupation as, "tanner, merchant, cordwainer, yeoman, cooper or cabinet maker". Sometimes they describe the house and outbuildings, roads, right-of-ways, and even the location of gardens and orchards. The more information, the easier it

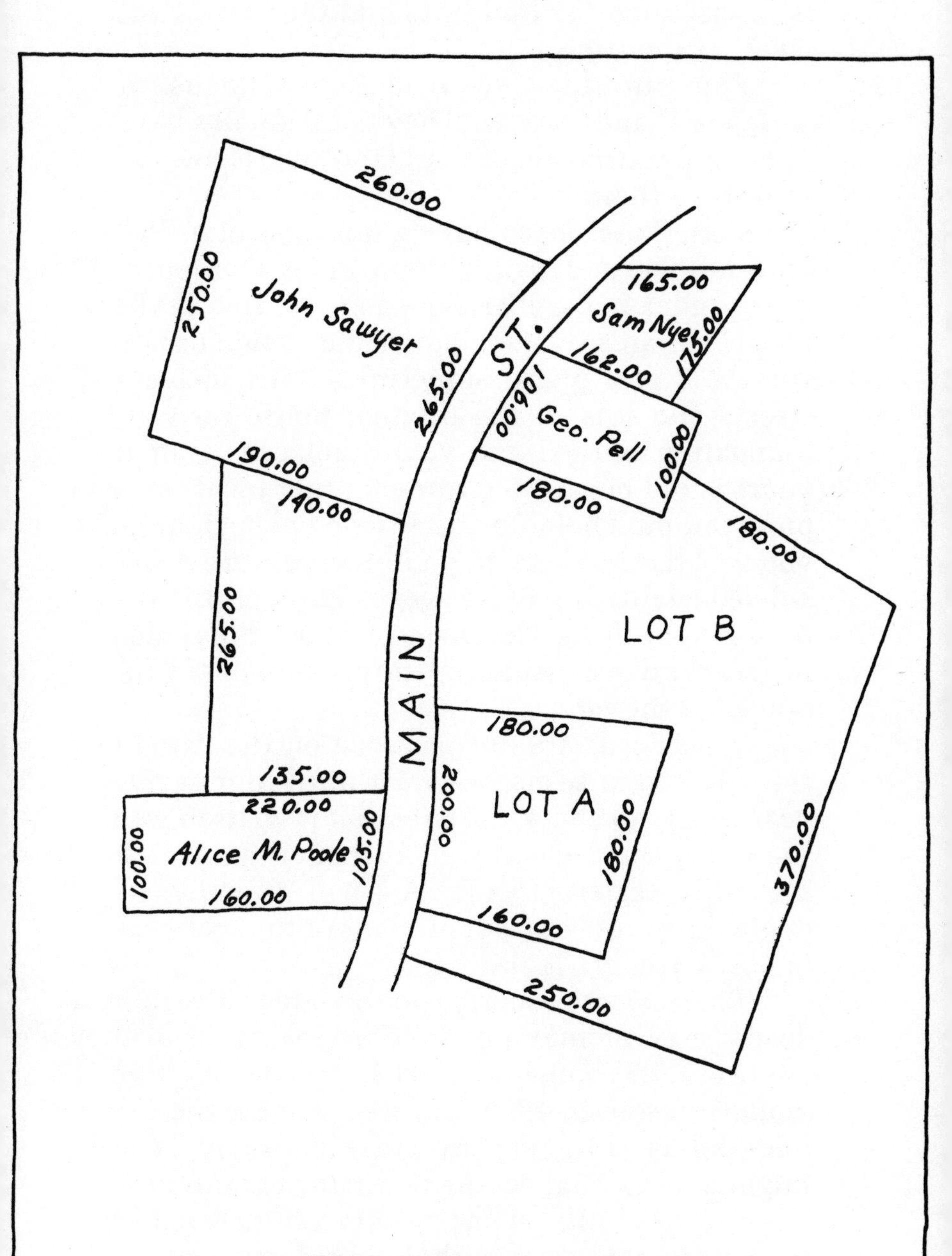

Plot plan sample

is to visualize the buildings and the "lay of the land" at the time.

You should also read some deeds of peripheral land owners. They occasionally have vital information about your land and its owners as well as theirs.

Some later deeds have a notation of a Plan Book number and page. This note may also appear in the index book. Plan Books are located in the Land Plan area of the Court House and contain surveyors plot plans, sometimes with adjacent streets and lots. If the original house on your foundation burned and yours replaced it, or if yours is old but moved from another location, a plot plan may help. In the absence of such help, you will have to rely on your own carefully considered evaluation of its age, to know if it is the house in the deed. The shape of your lot may also interest you in considering the placement of your house on the land.

Most deeds are not recorded on the date of the sale and in some cases not until four or five years later. So check the date that is written into the end of the deed, as well as that of the recording at the Registry, below. A gap of several years could throw off your calculations or the sequence of deeds you are looking for.

The cost of property, often stated in earlier deeds, may or may not hold a clue for you. If twenty acres bought in 1702, for twenty five pounds, resold in 1704 for forty pounds, the differential is sufficient to cause curiosity. You might assume that during that two year interval, a house was built on the twenty acres. A list of the acreages, with or without buildings, and the cost can be analyzed for disproportionate jumps in the value of your estate over the years. A cost increase or decrease, not attributable to fluctuation in land size, could mean the addition or subtraction of buildings or a major remodelling of your house.

The chart should include a breakdown of the price per acre. This can then be double checked against deeds of other property in your area for that year, to determine if it was the "going rate". If it was twice as high per acre as a neighbors, it may have included a building even though there is no mention of one in the deed. Once in a while, an old deed will not include a monetary value, but rather say, ". . . partly in consideration of the good affection I bear unto my beloved son and partly in consideration of sundry sums of money . . .". The sentiment is appealing, but it may come at a point in time when you would prefer pounds or dollars practicality. Another problem is that of making the distinction between inflation or appreciation, and the possible improvements to your house. The cost approach may not work for you, yet you should be aware of its potential.

It is advantageous, if possible, to make a copy of each deed to take home for further study and reference. It will avoid the anguish of forgetting to note an important detail and will save you the trouble of an extra trip to the Court House.

When the deed search is finished, you may have a good idea of the age of your house. You may feel, however, that your house is older than the oldest deed indicates. Maybe you are stuck at the transfer of the estate through a will or wills.

Wills

A last will and testament is a more personal document than a deed. While it may be cold and legal, it may also be the expression of warm human feelings of a dying person anxious to take care of his family's future.

Wills are kept in the Registry of Probate at your Court House. All original documents pertaining to a willed estate are bundled together in a case and put on file. In a Court House with limited space, wills may have been put on microfilm. If you are unfamiliar with the use of the viewer, hopefully someone will be there to help you. These papers, wills, inventories and other legal notes are typewritten, printed, or in the instance of older ones, often written by hand in a fine script. (An old will is delightful to look up and read, even if you do not need it.)

Again, this collection is referenced in an index. The indexes are arranged alphabetically, by time spans. Look for Richard Hall under H. in the volume or file cabinet containing the approximate year of his death, note the number and a clerk will find you the material.

Each case has a Docket number on the outside refering to a Docket Book. This book chronologically lists all legal proceedings of the Probate Court and short summaries of the will and other papers. It is regrettable that some early originals may be missing from a case; a particularly choice will or inventory. The Docket Book, however, should tell you what you want to know. There will also be a set of indexes for quick reference to Docket Books and other records.

It is easy to waste time looking up someone

in your Court House who died in another county. A will is always probated in the county in which the individual died, not in the one in which he spent most of his life, unless he perished there. This search may also be complicated by the ever present duplication of names. No matter how distinct you think the name, the ubiquitous double is always there!

Early wills are simple; designating house and land distribution and the sharing of household goods, to members of the family. An inventory, the list of a man's worldly possessions, can tell you his occupation and give clues to the extent of his estate, cataloguing livestock and buildings. "Barrels in the shed cellar", could alert you to an area in your basement that you know nothing about. A run-down of the heirs and their ages gives you an idea of the number of dependents living at home and hence a hint of the house size. A man, wealthy by the standards of the day, might have a large house, rather than a small one. Or, if he was young at his death, as was often the case, and he left everything to his wife, their home may have had one room and a loft. Read between the lines of a will and you can learn more than a cursory glance will yield.

One will may not satisfy your needs. Should the heir or heirs sell the house, you are all set, but if they lived on in the place and passed it on to their children, there may be enough passage of time to cause confusion.

Suppose that the last deed you can find shows that William Brown sold your house in 1890, but you feel certain the house is much older than that era. You have checked in the Grantee Index and no one sold any house to William

Brown. A check in the Will Index reveals that W. Brown inherited the place from his father, James, in 1850. The will also tells you that James was 91 years old. Now you know that possibly James Brown is the Grantee that you are looking for. Returning to the Deed Index, you now find that a James Brown bought the house in 1810 from his father, Charles. Charles Brown may or may not be the original owner, but at least you are back to the deed books again. In the 17th and 18th centuries, it was common practice for a man to sell his estate to his son, though that does not preclude the chance of his leaving it in a will.

There may be other complications, such as a man leaving his estate to his wife who remarries, (a change of name), or a married daughter, (another name change). Also an estate may be passed down through several generations, leaving a huge gap of years to be accounted for by wills.

A will may be probated at the death of a land owner, with the property held by the family and rented for a number of years before it is sold. Similarly, a man may buy an estate for the purpose of using the land only, while leasing the house.

Old wills are fun to read and can be exciting emotional documents, revealing the personality of someone who once lived in your house.

Town History and Genealogy

Many towns and cities have a published history, written by a public spirited citizen. Although these books are apt to be out of print, they will be available at the library and some lucky individuals have private copies. A genealogy, as part of the history or separate, is a great help. These books tend to be old, but some have been brought up to date.

A local history may list your house in a street by street description of the town. It might even state the name of the original owner. Sometimes those in residence at the time of publication are mentioned and a rough description of the house to identify it on the street. Prominent men are always given lots of ink and are chronicled from the settling or incorporation of the town. "Ner-do-wells" can also come in for notoriety. You may discover that selectmen or Civil War soldiers were among the previous owners of your home.

Genealogies are interesting for their information on the size of families. If you find that the builder of your house had twelve children, one born each year, you can assume that your place was sizable early on. Unless, of course, they slept six to a room! A student of architecture can look for structural evidence in his house that would reflect the regional differences of the country its builder emigrated from. Many ideas can be extrapolated from the genealogy of an early settler.

Published and Unpublished Books, Articles, Memoires

Town libraries are a source of wisdom, researched or recalled by local citizens. Many times, old people have been urged "to write it down before you forget it", and the result is a

YOUR TOWN
1850

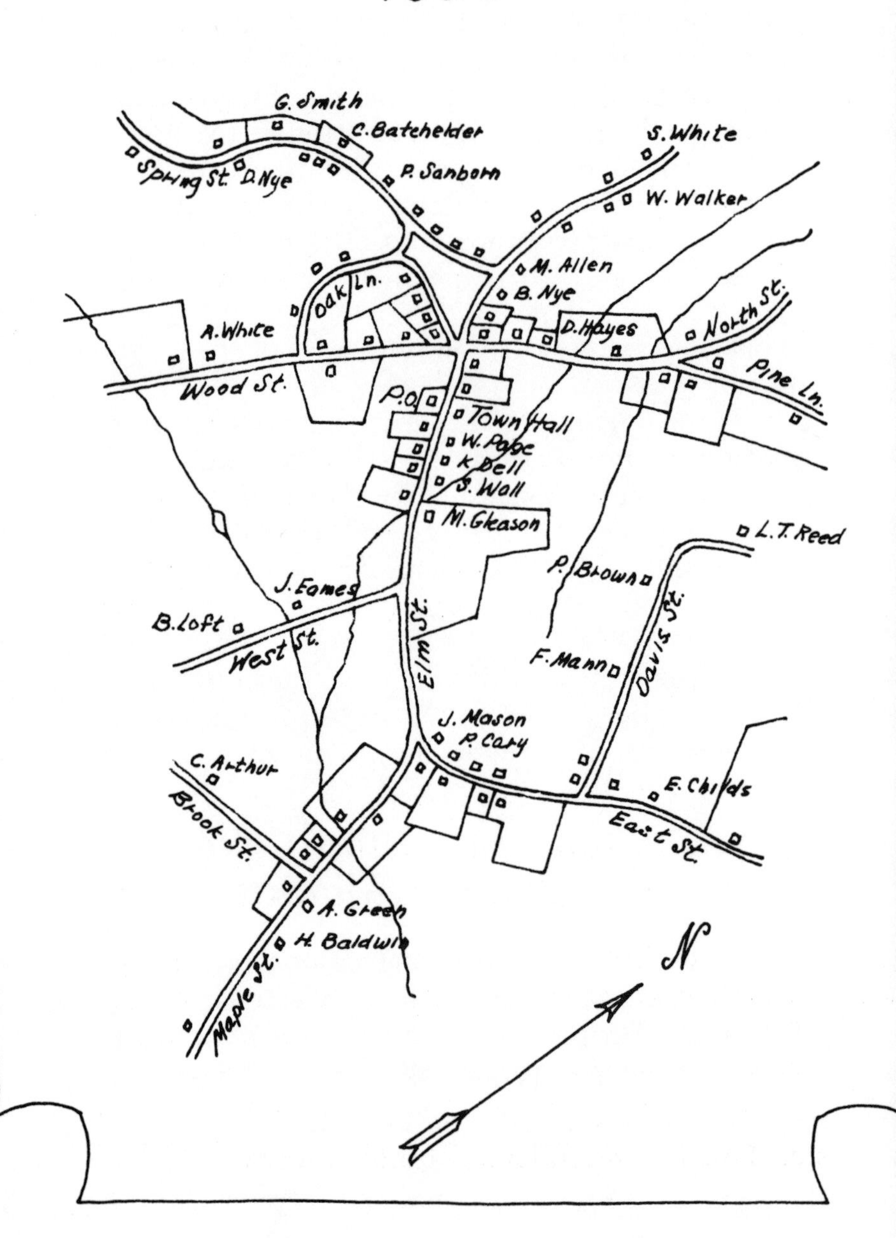

charming remembrance of details and insights. This "fount" supplements and may surpass portions of a more formal history. It also may contradict that history. Again, you could encounter a street by street exposition of your area. The "I remember when", or "my grandfather told me", might contain a useful reference to your house or the people who lived there.

Maps

All local maps which you can find at the library or elsewhere, have the potential for defining old boundaries and the position of your house. Keep in mind that names of streets change over the years, as do house numbers.

By piecing together facts garnered from deeds and sections of old and new maps, you can draw yourself an approximate plot plan. The configuration of land or original acreage and the placement of your house on it, might influence your thinking on its age. Suppose that the house sits on land composed of two lots, each outlined in a separate deed stipulating a dwelling-house, and they are dated years apart, then you must decide which deed to believe. A plot plan of that time may help.

A local historian may not only have done some writing, but tried his hand at map making, as well, with early houses marked on original King's Grants. Or you may find an effort to break down the grants to a land map of later home owners with your house drawn in and a tentative date, maybe even a name. An amateur job is not always trustworthy, so check it carefully, before you take it as gospel truth.

A deed with boundary markers of "elm trees", "heaps of stones" or "ditches" is admittedly difficult to work with, though not always impossible. It may tax your ingenuity, but modern land areas are often clean-cut subdivisions of earlier farms and these in turn of larger holdings.

Town Register

At some point in time, your town probably began to publish an annual register of all its residents. The first ones will give the name of the street and an imprecise location for your house, like "opposite Perry Road" or "near the road to Wells mill". Beware, again, of later issues with street names and house number differences. Your "Elm Street" may have been "Main" in the 19th century. And number 382 may now be 1220. This is something to keep in mind, too, while reading deeds.

(Inevitably, as you make your way through local literature, you will become acquainted enough with the growth of your town to assimilate street descriptions and changes. The adjustment in street layouts and name and number changes will be briefly noted in one source or another.)

Renting or loaning a house has always been a common practice, as was "letting" to boarders. Your house may have been lived in at one time, by a succession of people who never owned it. Hence, early registers may show people in your house for whom you can find no deed. Or they may give you that one badly needed name for which you have been hunting.

Historical Society

Do you have an Historical Society in your town? These organizations are a traditional repository of gifts and bequests of valuable research material. Documents, original deeds, maps, pictures, letters, often find their way into societies for their preservation. Contact the archivist or historian and explain your problem and maybe your worries will be over. It is the researchers dream that the one deed that failed to be recorded in a Deed Book, will eventually turn up alive and well in its own home town. Pictures that have been forgotten for decades, may surface and be tucked away again in your Historical Society's files. Urge the archivist to dig deep for all photos and etchings.

Society proceedings can also be a source. Sometimes they are published yearly, with reprints of lectures on the history of the town. "Yesterdays" are a favorite topic for such talks and may go far enough back to be of service.

Assessors and Tax Collectors Office

Land has been valuated and taxed since the beginning of time. Although the records on your town may not go back that far, they are another source.

Real estate, buildings, horses, cows, swine, sheep; none escape the notice of your friendly tax collector. Tax records fix people in a town at a date you may not easily obtain elsewhere. And the extent of their holdings may be of more than passing interest.

In the same way that an unusual increase in the price of your place, from one deed to the next, may intimate a new house or improvements, a large tax increase from one year to another may also be significant. If you think your kitchen ell was added on sometime in the late 1800's, the tax records of that era, if available, may verify your suspicions.

Books and Articles on Old Houses

Interest in the old and antique is at such a peak today, that there are innumerable excellent books and articles about old houses. Although specific dates of architectural change are difficult to come by, you may chance on a detail in a picture of an old house with a known date, that corresponds to a similar part of your home.

House shapes, floor plans, details of beam construction, size and form, or the composition of plaster may inspire you. Information of this sort is always helpful as a guide to antiquity.

A study of authoritative publications can only add to your repetoire and will sharpen your skill at recognizing prominent features of age. The selected book list at the end of this guide will get you started.

V WHAT DOES IT ALL MEAN?

Your hardest work is now over or just beginning. If you were unable to zero in on that mysteriously difficult date, now is the time to cogitate and trouble shoot.

Try different ideas. Put your material together in alternate patterns. If it is a boundary problem, go out doors and look for a granite marker. Double check all architectural details. Exercise your imagination and try to visualize your house and its setting, "way back when". Think about how your house grew and the sensible way for it to start. Pretend that you were the settler that did the building and farmed the land. How did his relatives fit into the picture?

Maybe you made a wrong turn or missed a deed. Go back over your data and study it some more; re-think it. Sometimes it helps to type out passages of deeds that are difficult to read. Not all early recorders got an "A" in penmanship! The legal wording can be confusing and takes time getting accustomed to. If you let your mind wander and consider ridiculous possibilities, something new and obvious may pop up.

Scream for help in interpreting your findings. Someone might hear you and come to your aid. We all can get a blind spot and miss the simplest solutions. There may be an expert near by who would enjoy seeing your house at the expense of helping you psych out its age.

All of this should bring you within a few years of the birth of your house, if not closer. Then it is anyone's guess, but you will be the authority and should feel that for your reward, you can pick a date and defend it.

VI CASE STUDY: THE JOHN MASON HOUSE

BUILT 1680
RANSACKED BY THE BRITISH
PROBABLY THE OLDEST HOUSE
IN TOWN
HERE LIVED JOHN MASON

When a plaque like this on a house meets your eye, what do you think? Most people read it, believe it, and are impressed. The date may or may not be authentic, but the average house-looker is not discriminating; an old house is an old house. When you stop to think about it, very few houses of that age are left and fewer still retain many original features. Yet, most of us will wait to be convinced and even the wording on the sign will not worry us. Although "probably" is a word that does not convey much assurance, we will look the house over and go inside to walk around, still captive. It is human nature for the proud owner of a fine colonial to always pick the oldest date, when in doubt. If there are other old houses in the neighborhood, he may fudge a little to make sure his is the oldest. Such was the house that I moved into.

The previous owners had generously left behind a small library of books, including a town history, Historical Society publications and a scrap book of pictures and words about the house. One set of pictures revealed that the same plaque had been on the house through several owners. Only the last had tentatively dared to challenge, in a brief history, the accuracy of the 1680 date on the outside. I made a mental note of this concern, but a number of years went by before I was sufficiently stimulated to do anything about it.

When a local group agreed to take on the job of dating all the old houses in town, they asked me if I would do a history of the John Mason house. I accepted cheerfully and then sat down and wondered how to do it. Oh, for a guide book! Typically I roared off in many directions at once; read a little here and a little there, and assimilated nothing of what I had found. I recalled that someone, in the dim past, had mentioned visiting the County Court House to research her deeds, but I had neglected, at the time, to take any interest in her story. Finally, I settled down, dredged up the material from our lawyers title search and went to work in earnest.

The lawyer's findings had produced information back to 1910 and although I did not realize it at the time, I was immediately plunged into a dead end. The last solid document I had was part of a Mr. Munroe's will, authorizing the sale, by executors, of his estate. Braving the court house Registry of Deeds. I spent hours of fruitless research, trying to match the Grantee, James S. Munroe with his Grantor, the man from whom he would have bought the house. It did not occur to me that it would be difficult working from a will to a deed and what I did not know and could not guess, was the length of time that he owned the house.

I found some unpublished history in the local library; three volumes in manuscript form of rambling and sometimes contradictory research and memoires. This lengthy work proved invaluably rich with information, but I missed the most important names I was searching for. In the meantime, I discovered a collection of Town Registers. Divining that my house was "North of Main Street, near Munroe Station", I soon compiled a long list of inhabitants and rushed back to the Registry of Deeds. None of these names, people who had lived in my house from 1898 to 1916, were to be found in any index.

Returning to the library, I re-read whole sections of the unpublished history and finally discovered that William Winthrop had owned the house in 1795. This time I made copies of all the pertinent pages and took them home for further study. Off to the court house again, I triumphed at last. William Winthrop did indeed own the house from 1791 to 1822, when he sold it to Jonas Munroe. Time, always short, ran out and unable to get past Jones, I returned home with a copy of only one deed.

Turning now to another source, I found in the town history and genealogy that William H. Munroe was heir to Jonas's fortunes. I read, too, with avid interest, the genealogy and short sketches on John Mason and his family, anxious to place them in my house. My next trip to the Registry produced a deed from William H. Munroe to James S. Munroe. No wonder I had trouble back tracking from James S. He had owned my house for forty eight years, renting it out nearly the whole time.

Now I was free to collect the last of the deeds, those of the builder, John Mason and his sons, John Jr. and Thaddeus. I found a couple of indenture deeds, as well, including the one in which the Mason family forfieted the estate for lack of mortgage payments, after a continuous occupancy of more than ninety years. I still lacked the

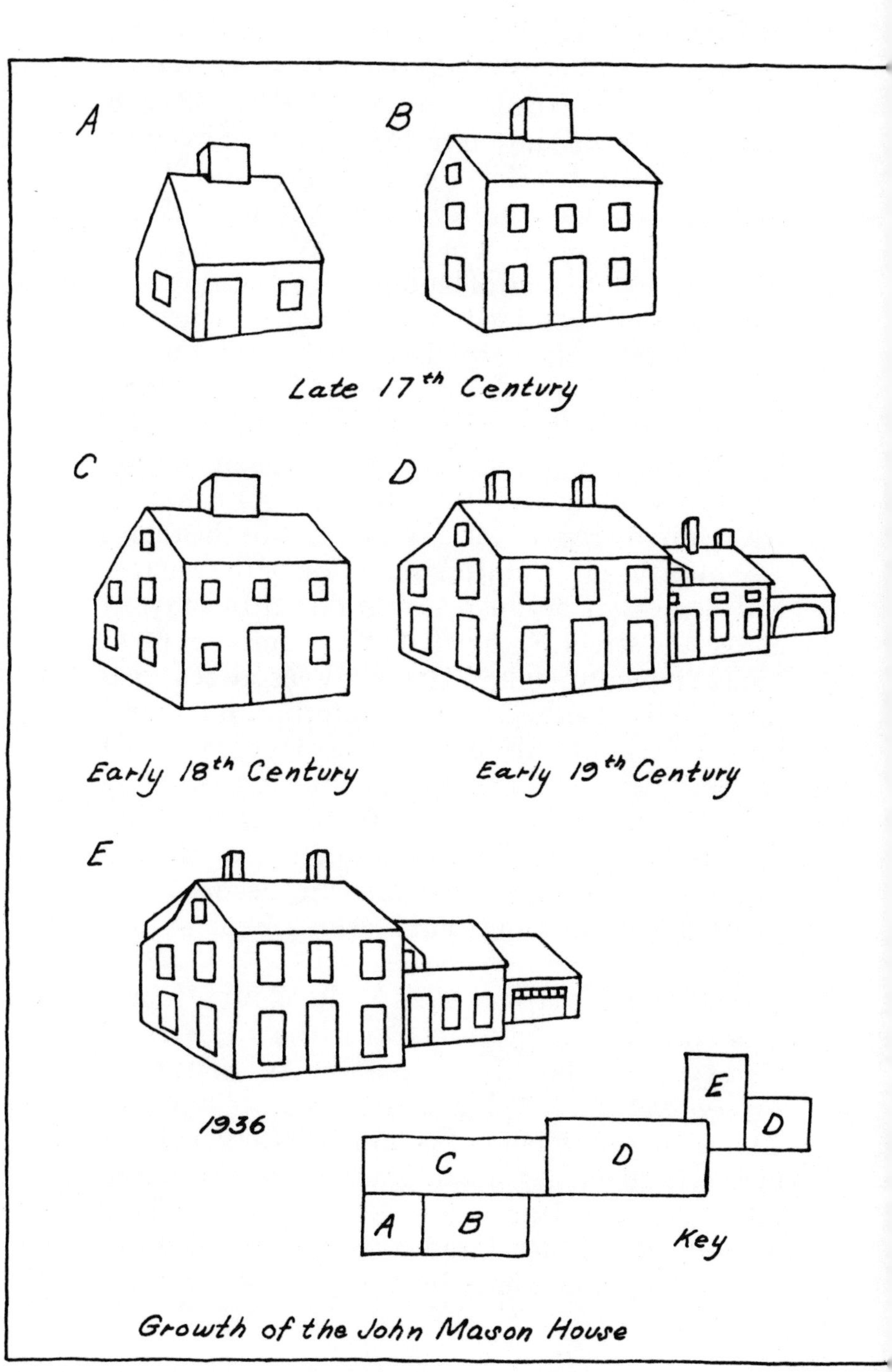

Growth of the John Mason House

one deed I needed, however, and that was for proof of John Mason's right to the land. The Town History gave me the information that he had been admitted to the church in 1699 and that his first son was born in 1700, which lead me to believe that he must have built himself a house by that time. A stronger proof of his occupancy was contained in a 1699 deed of land adjacent, stating that John Mason possessed land formerly owned by Joseph Estabrook. Joseph Estabrook was his uncle and several of my sources speculated that John had somehow "got" the land from him.

My search was ended. With reasonable assurance, I put down 1699 as the date of my house, wrote up the report and submitted it to the team.

Months later, at the suggestion of the Historical Society, I began research on an old tavern across the street. Somewhat more confident now, I made fewer mistakes. Using my resources with more skill, I found all the information that I needed except for the deed of one owner in 1693. I wrote up the story of the tavern and as it lay collecting dust awaiting the final polish for publication, I became more and more tantalized and curious about those two missing deeds. Especially the one proving John Mason's ownership of my own land.

Once again, I was back at the Court House, this time to try my luck with the Registry of Probate. I explored the possibility that John Mason had inherited the land from his uncle or some other member of his family. No luck. Back at the deeds again, I was stunned to find eight more deeds pertaining to Mason family land transactions, which I had totally overlooked before. How could I have been so dumb? From the late 1600's through most of the 1700's, there are so many John and other Masons listed in the index, that I had been reluctant to spend the time needed to track them all down. Time, in the court house passes by quicker than the Tokiado Express. It is always time to go home!

Armed with copies of these new deeds, I hurried home to read them. Now, I found that I had a deed to six acres, called "demised premises", of Mason land, bought from John's Uncle and also one for an additional twenty acres bought from a Mr. Emerson. The six acres were purchased in 1706. But, how could that be? The deed for the land across the street showed John Mason on those six acres back on 1699. To further confuse me, the twenty acres, bought in 1703, showed John Mason on the "noreasterly" boundary and of all things, listed a dwelling house!

At this point, I became very excited, nearly irrational. Could the 1680 date on the house be possible after all? A house on Mr. Emersons twenty acres might be ancient. I took my flashlight and looked the house over again, inch by inch, trying to make it older than I really felt it was. On a trip to the library, I found and brought home all the books I could carry with details of house construction. I tried to locate Mr. Emerson in the Town Genealogy. No information. It occured to me to check the deed for the land when he purchased it, to see if there was a house. The deed book was missing from the stacks and I was directed to a room where a reproduction of it was being corrected. The book was out of its bindings, lying in piles on a filing cabinet. The women in charge, after a lengthy consultation, decided that I could copy the deed, but that one of them must accompany me. All but handcuffed, I was watched with a sharp eye to make sure that I did not run off with that precious document. Back at home, I found that I had a copy of the first page only. I still needed the date at the end as well as other deeds from that book. When I returned, I didn't have the nerve to go through the kidnapping process again, so I just went to the room and asked for the date, leaving the other deeds until some later time. The deed itself, was for *one hundred* and twenty acres and included a house. Obviously more research is required, but I shy away from the charade required to do it.

Also, I put together a composite map of all the pieces of acreage owned by John Mason, trying to decide if it was possible for the house to be on the twenty acre lot. Across the street, I found a section of an old stone wall, hidden in the shrubbery, in exactly the right place to support this idea. I took my time looking up the definition of "demised".

Finally, I found some more deeds showing that Mr. Emerson had sold off other parcels of his estate and each of these deeds had the same wording as mine, including mention of a house. I had to acknowledge that it was unlikly Mr. Emerson would have three houses in the very early 1700's, one for each plot of land that he sold. I can not explain the stone wall in a perfect place for an old boundary, except that is must be a later addition. And I admit that the wording, "demised premises", in Uncle Estabrook's deed, must mean that John Mason leased land from his uncle to build his house. There is no indication of how long he may have rented this land, but I feel that 1699 or 1698 would be a reasonable date for the house.

Still plagued by lack of final proof, I am never-the-less, resting my case for the moment. In keeping with tradition, I have chosen the older date and if I am challenged, I feel that the data almost supports my claim. But, I do not expect to be tripped up because the plaque on the outside of my house now reads:

BUILT 1698
RANSACKED BY THE BRITISH
ONE OF THE OLDEST HOUSES
IN TOWN
HERE LIVED JOHN MASON

GOOD LUCK

VII POST SCRIPT: WRITE YOUR OWN REPORT

Historic Commissions in many parts of the country are pushing for a comprehensive survey of all buildings constructed prior to 1900, or there abouts, as well as outstanding examples of newer structures. Some day, someone might ring your bell, if they have not already, to solicit your help. You may want to put the results of your research into a readable form in anticipation, or for your own gratification. And, too, your newly developed expertise may tease you into volunteering to research someone elses house. Detective work of this nature is badly needed in many towns recently awakened to the need for historic documentation.

A chronological list of dates and owners, with deeds, wills, and renters, is a good beginning. It is also fun, (you do not have to be a talented writer), to summarize your findings, with commentary about the lives of past owners. The inclusion of doubts and questions may spur you on to further study at a later date or inspire some future owner to take up the battle.

Consider also, slipping some copies of your work into covers and passing them out to your library and Historical Society. You will then become a resource for others or provide novel reading for local history buffs.

In any case, whether anyone cares or not, you can enjoy the satisfaction of knowing that a job-well-done is behind you. I bet someone will care.

VIII GLOSSARY OF DEED TERMS

ABIDE AND REMAIN - continue.

ADMINS. - administrators.

AFORESAID - already mentioned.

ALIEN - give up.

ANY WISE - any way.

APPERTAINING - pertaining to, belonging to.

APPURTENANCES - incidental property rights or privileges.

ASSIGNS, ASSIGNES - those to whom a right or property is legally transfered.

BARGAINED PREMISES - property to be sold.

BE THE SAME - are the same.

BENEFIT AND BEHOOF - advantage.

BESIDES - plus.

BUTTS AND BOUNDS - abuttals and boundaries of a property.

CONFIRM - validate.

CONVEY - transfer.

COVENANT - pledge.

CULTURES - tillage, crops.

DEFAULT - neglect to pay.

DEMISED PREMISES - under English law; leased land.

DIVERS - several, various.

DO - ditto.

DO HEREBY - do by this; do by means of

EASEMENT - deeded and limited use of another's land, as the right to cross over the land; right of way.

EGRESS - inlet; way into.

EMITTING - giving out.

EMOLUMENTS - benefits.

ENCUMBRANCE, INCUMBRANCE - a claim or lien upon an estate.

ENSEALING - impress with a seal.

ENTAILS, INTAILS - restrictions of inheritance settlement.

EXCEPTING - but; except.

EXEC^{TOR} - executor; a person appointed by the deceased to execute the will, to see its provisions carried into effect.

FFARM - farm.

FFEE SIMPLE, FEE SIMPLE - under English law; means by which land may be inherited by any heirs of the owner.

GOOD AND LAWFUL BILLS - legal money.

GRANTED PREMISES - sold land; land about to be sold.

GRANTEE - person buying house or land.

GRANTOR - person selling house or land.

HERBIAGE, HEARBAGE, HERBAGE - pasture.

HEREOF - of this.

INCUMBRANCE - see encumbrance.

INDENTURE - mortgage.

INSTRUMENT - deed of mortgage or sale.

INTAILS - see entails.

JOYNTURE, JOINTURE - an estate settled upon a wife.

JUSTLY INDEPTED - legally in debt to; fairly in debt to.

LAWFULLY SIEZED AND POSSESSED - legally owned.

LAWFUL SUIT - law suit.

LETT, LET - hindrance.

LITTLE MORE OR LESS - approximately.

LYETH - lies.

MADE OATH - swore; attested.

MANIFESTING - freely giving.

MANNER FOLLOWING - the following way.

MANSION HOUSE - dwelling house, large or small.

MEADOW - low land.

MEASURE - amount of land.

MESSUAGE - dwelling house.

MOLESTATION - interference with or troubling another in his possession of land.

M^R - Mr.

NOREASTERLY - northeasterly.

NORWESTERLY - northwesterly.

POLE - rod; 16½ feet.

POWER OF THIRDS - a widows dower or right to one third of her husbands estate.

PREMISSES, PREMISES - property.

PROFFITTS, PROFITS - advantages.

RAIL - fence.

RATE - interest rate.

REMISE - give up claim to.

RIGHT OF DOWER - brides portion.

ROD - 16½ feet.

ROGRESS - outlet; way out of.

S^D - said; already mentioned.

SEIZED - owning.

SETT HANDS, SET HANDS - signed.

SITUATE, BEING SITUATE - located.

SOLE AND PROPER OWNER - only legal owner.

SUCH MANNER AS - like.

SUCH OTHERS - any others.

SUNDRY - various, miscellaneous.

TENEMENT - dwelling house.

TENOUR - tenor, intent.

TESTATE - having made and left a will.

THESE PRESENTS - witnessers and signers of the deed.

THRO - through.

TO ME IN HAND - given to me.

TO WIT - namely.

TRACT OR PARCEL OF LAND - piece of land; acreage.

UPLAND - high land.

VIZ - namely.

VOID - cancelled.

WARRANT - guarantee to protect.

WATERCOURSE - natural or man made channel.

WAY - highway or road.

WITTNESSETH, WITNESSETH - witnessed.

W^{TH} - with.

Y^{T} - that.

Y^{E} - ye.

& - and.

IX LIST OF OLD HOUSE BOOKS

Arthur, Eric and Witney, Dudley. *The Barn.* Ontario, Canada: M.F. Feheley Arts Co. Ltd., 1972.

Bruce, Curt and Grossman, Jill. *Revelations of New England Architecture.* New York: Grossman Publications, 1975.

Bullock, Orin M. Jr. *The Restoration Manual.* Norwalk, Connecticut: Silvermine Publications, Inc., 1966.

Chamberlain, Samuel. *A Small House In The Sun.* New York: Hastings House, 1936.

Hamlin, Talbot. *Greek Revival Architecture In America.* New York: Dover Publications, 1964.

Isham, Norman Morrison. *Early American Houses And A Glossary Of Colonial Architectural Terms.* New York: Da Capo Press, 1967.

Jackson, Joseph. *American Colonial Architecture.* Washington Square, Philadelphia: David McKay Co., 1924.

Kelly, J. Fredrick. *Early Domestic Architecture Of Connecticut.* New York: Dover Publications, Inc., 1963.

Kauffman, Henry J. *The American Farmhouse.* New York: Hawthorn Books, Inc., 1975.

Maas, John. *The Gingerbread Age.* New York: Rinehart, 1957.

Mercer, Dr. Henry C. *The Dating Of Old Houses.* From The Bucks County Historical Papers, Vol. V., 1923.

Morrison, Hugh Sinclair. *Early American Architecture, From The First Colonial Settlements To The National Period.* New York: Oxford University Press, 1952.

Pierson, William Harney. *American Buildings And Their Architects: The Colonial And Neo Classical Styles;* Vol. I Garden City; New York: Doubleday & Co., 1970.

Sloan, Eric. *A Museum Of Early American Tools.* New York: Ballantine Books, 1973.

Whiffen, Marcus. *American Architecture Since 1780: A Guide To The Styles.* Cambridge, Mass. M.I.T. Press, 1969.

Williams, Henry Lionel and Ottalie K. *Old American Houses 1700 - 1850.* New York: Bonanza Books, 1967.

Williams, Henry Lionel and Ottalie K. *A Guide To Old American Houses 1700 - 1900.* South Brunswick; New York; London, 1977.